A Random Act

*An Inspiring True Story of
Fighting to Survive and
Choosing to Forgive*

CINDI BROADDUS

with Kimberly Lohman Suiters

Harper

An Imprint of HarperCollins*Publishers*

A hardcover edition of this book was published in 2005 by William Morrow, an imprint of HarperCollins Publishers.

HarperCollins books may be purchased for educational, business, or sales promotional use. For information please write: Special Markets Department, HarperCollins Publishers, 10 East 53rd Street, New York, NY 10022.

First Harper paperback published 2006.

Designed by Katy Riegel
All photographs courtesy of Cindi Broaddus

The Library of Congress has catalogued the hardcover edition as follows:

Broaddus, Cindi.
    A random act: an inspiring true story of fighting to survive and choosing to forgive / Cindi Broaddus with Kimberly Lohman Suiters.— 1st ed.
       p. cm.
    ISBN 0-06-073514-7
    1. Broaddus, Cindi.   2. Chemical burns—Patients—Biography.   3. Burns and scalds—Patients—Biography.   I. Suiters, Kimberly Lohman.   II. Title.
RD96.45.B76 2005

617.1'106'092—dc22
[B]   2004061865

ISBN-10: 0-06-073515-5 (pbk.)
ISBN-13: 978-0-06-073515-9 (pbk.)

06 07 08 09 10 WBC/RRD 10 9 8 7 6 5 4 3 2 1

This book is for Angela, Shelli, and Brandi
for all that you are and everything
that I am because of you

Contents

Foreword   ix

1. Am I Dying?   1

2. Gathering My Army   17

3. I Am Not a Victim   29

4. Revenge Is Not Mine   49

5. Humor Doesn't Hurt   71

6. Going Home   87

7. Facing Fears   109

8. In the Spotlight   131

9. Lawmen and Lawmakers   159

10. Random Acts of Kindness   179

Epilogue   203
Acknowledgments   207

Foreword

by Dr. Phil McGraw

I have always said that the decisions we make each day create the outcomes in our lives. If a vile, heinous crime were to ravage you and your family, if tragedy were to knock on your door, how would you react? Would you shrink into yourself, fearing life and constantly looking over your shoulder, or would you step up and challenge the label "victim"?

You are about to hear a story that will make you stop and think. It is an amazing story of courage in the face of tragedy. Not long ago, a jar of acid was thrown from a highway overpass and crashed through my sister-in-law's car, burning her all over her body. This senseless, random act of violence by a stranger changed Cindi's life forever. She suffered third-degree burns, countless surgeries, and disfigurement.

I am still struggling to get my mind around such a stupid and random act. Why, out of all the people in the world and in all the minutes in time, why did *she* have to be under that bridge, at that moment, on that road, when that sick and demented monster dropped that acid? How could such a horrendous thing happen to such a good person? Because I know Cindi, because she is family, I am sure that I am by no means objective. So let me just tell you the facts, facts that I believe will make you as proud and humbled to know her through this book as I am to know her as a friend and family member.

As a single mother, Cindi did whatever she could to make ends meet. Working three or four jobs at a time, she put her three daughters through college, never asking for or accepting anything she did not earn. She never complained, she never seemed to get tired, she just did. She leaned on God's strength and wisdom and put her family first, always doing what she had to do for her family, not merely to survive but to *thrive* with love and joy.

I have always believed and taught that *You Create Your Own Experience.* Where others might understandably find despair and frustration, Cindi has found strength, inspiration, and faith. Though she has and will continue to face a lifetime of surgical and rehabilitative procedures, and the pain and loss of function that come with those challenges, Cindi has chosen not to let that assailant hurt her again and again by robbing her of her spirit and optimism. Cindi could have chosen to wake up every day bitter and angry, but instead she chooses to wake up every single day and see the blessings in her life.

This book is the compelling story of how Cindi has found and embraced the strength to overcome. She still wears the outward scars of that day, but those scars have not changed the fact that she is truly beautiful both inside and out. She radiates good humor to all those around her. She does not look back in anger; instead she looks forward in joy to her family, friends, and colleagues, all of whom have supported her in her darkest hour. Her remarkable spirit has soared in the face of adversity. It is a spirit that is wonderfully contagious to all those who know her or read about her.

It's easy to be a victim, but Cindi's never taken that attitude. She has instead made the choice to be much more than a casualty of her attacker's madness—she has challenged herself to move beyond her circumstance. In this book she uses the understanding of her own experience to create a path, a plan for others to learn and, ultimately, to find their own way, whatever their challenges may be.

I am so proud of my sister-in-law, have so much respect for her, for her integrity and her selfless desire to help others, and I believe that you will, too. This book is as powerful as it is important. Read it with your family, look for strength in one another, and find your hero.

Cindi Broaddus is certainly a hero of mine.

Me, on my last birthday before the attack

1

Am I Dying?

I NEVER HEAR it coming. I never see it fall. And I never hear the bloodcurdling screams that come from my own mouth.

I do not feel anything.

I must be dead.

But I can't be dead because I am aware that I have a few seconds left. So have I come back to life? Or was I never dead at all?

The only explanation is that I am suspended in a void between life and death. It is a reprieve, a few seconds to tell my children good-bye.

My first conscious thought is that Jim, my friend who is driving, will have to carry that message for me. I hear myself begging him.

"Jim, please, pull the car over. Something has happened to me. I am dying. You have to tell my girls how much I love them. Tell them I'm so sorry, I don't want to leave them like this. Please pull the car over, Jim, please."

I need him to listen to me, to give me every ounce of attention he has, but why won't he stop the damn car? The darkness is overwhelming, not only outside in the predawn blackness, but also inside of me.

Though the fire is scorching my skin, I can't get a more searing image off my mind: my three daughters. The more I slip into the fiery abyss, the farther away from them I feel. *I cannot die like this. I have too much living I want to do. Please, God, get me home to my family. This cannot be real. I don't want Angela, Shelli, and Brandi to know I died like this—in so much pain and so afraid. I just want them to know I love them so much. I would give my life for each of them. We have been through it all together. If this is my end, I don't want my last moments to be apart from them.*

It's so dark and loud that I'm afraid Jim won't hear my desperate message unless he stops the car and listens to my every word. I need him to pull over so he can be my eyes, my ears, my mouthpiece. Does he even realize what's going on? I must be dying the way my mother did. Of a silent, deadly heart attack. Her death devastated me, so I know how my daughters are going to feel. I can't stand that thought. I can't stand leaving without them knowing that my last thoughts are of them.

"I can't see! Jim, I cannot see. I'm dying. I'm dying! You must stop and listen to me. Please pull over, Jim. Please, you've got to pull over."

Jim hasn't answered me, hasn't explained why he won't pull over. Instead of Jim's voice, I hear the voices of my daughters. I see their faces, but I can't touch them. I need them so badly right now, and I know they'd rush to my side if they knew what I was going through while they were sleeping. Can Jim even hear me, or have I already died? Is my time up?

I'm screaming bloody murder, but to me, my voice is calm. It seems to have no effect on Jim. He drives faster. I can't see the road, the speedometer, or him, but I can feel him accelerating, regardless of my pain, my fear, my death. My mind is going a million miles per hour. I'm reliving my life.

"Shut up, Cindi! I won't do it. You're going to live to tell them yourself." Jim fires back with a force I have never heard from him before. He would never tell me to shut up unless it were a last resort to shake me out of hysteria. It works. His intensity is welcome because it tells me I am alive. *Oh my God, I am alive.*

But for how long?

That void, not knowing whether I was alive or dead, is gone. What replaces it is even worse. It is pain and fear.

"I can't pull over, Cindi. I have to get help. God help us, I think someone threw acid on us."

In the dead of night, on a long, lonely highway, someone crept along an overpass with a gallon glass jar of sulfuric acid, and waited. It was 3:00 A.M. on Tuesday, June 5, 2001. At that hour, the H. E. Bailey Turnpike outside of Newcastle, Oklahoma, might as well have been the bleakest, loneliest dirt road in America. Little did we know that someone else was awake at this hour, planning his attack, his heart filled with anger, his hands bent on destruction, and his eyes focused on the highway below. He had the upper hand as he waited for his defenseless victims, two innocent people who trusted that a quiet night meant a peaceful one. Two innocent people whose lives would be changed forever.

Did I hear Jim right? Acid? Who would do such a thing? But because Jim won't stop, and because of the desperation in his voice, I'm convinced someone is chasing us. Someone set me on fire and Jim is not stopping because now that person is after us. *Who would chase us? Why does he want to kill us? Where is he now, behind us, or right beside us? How do we get away from him?*

These answers can only come from Jim because I can't see. I have been blinded by the acid. My face catches fire first, then my neck and chest and arms and stomach and legs. The pain is unbearable. I've got to get it off me. With both hands, I cover my face, feel something moist, and believe I'm bleeding. But it's my skin coming off in my hands. My eyelids, my chin, my lips are all being eaten by acid.

Cautiously I take one hand away from my face, making sure the other is still covering as much of it as I can reach. I grab at my shirt and it shreds in my hand. My clothes are so soaked in acid that they burn my skin more deeply. I can't get my bra off, and my skin melts around it. I thrash around like a wild woman with no thought, just a feeling that I have to get this liquid fire off my body. It is eating me alive.

It seems like hours, but only a few minutes pass, when I finally hear Jim say the words I've been praying for.

"Cindi, I see lights. There's help up ahead."

I know that's what he said, even though the wind is deafening. The jar of acid had crashed through the windshield of my car, leaving an enormous gaping hole through which the air beats mercilessly against my face. The dark morning is cool, but it tortures my burning skin. A million tiny shards of glass pierce my cheeks, my arms, my breasts, and my legs. *Is this what dying feels like?*

After driving one and a half miles that feel like one hundred and fifty, Jim pulls over. He stops the car so abruptly I hear the tires screech. Because I'm literally blind, I can't see the horror on his face, but I'm about to find out why I'm going to die.

"I couldn't pull over until now. I'm sorry, baby."

Along the way, he'd spotted only the yard lights of distant farmhouses. We'd traveled this turnpike many times before and knew there aren't many exits. We'd never gotten off before reaching the city. Jim had driven to the nearest exit and turned toward the first bright lights he could see, about a quarter mile away. It's the Newcastle Indian gaming complex.

"Hold on, baby, I'm going to get help."

Without waiting for me to respond, he's gone. As the door slams shut behind him, another excruciating blast of air adds to my agony. For all that I can't see, I can hear Jim's footsteps running away, and then the quiet—the damn, deafening quiet. I once again feel the paranoia that someone is chasing me, Jim's abandoned me, and someone's going to get me. *Death is going to get me.*

My fear is suffocating. The abyss is black and spinning and sucking me down. If only I could see. I have to escape. I have to get out of this car so I can breathe.

I will breathe again. I will not be trapped. I will not sit here in acid and glass and shredded skin and let this take me. I'm about to make my first choice in this dire situation. I have decided to fight. I've been alone before, and I know I have to help myself before anyone else can help me. I could sit here, feel sorry for myself, and accept what's been handed to me or find out what's beyond this car door.

My right hand feels its way to the latch. I open the door and pull myself out, wondering where Jim is. It's eerily silent. No cars, no people, not even the ever-whipping Oklahoma wind. It's not what I expect. *No one* is rushing to help me. My moment of courage is fading quickly. I shut the car door. With my right hand again covering my eyes, I feel my way along the car with my left hand. I sense the heat of the engine. I have no idea where I am. I've never felt so alone.

I don't know where the hell I'm going, but I'm going. For a second, I take one tiny step too many and suddenly find myself panic-stricken. I've lost contact with the car, and fear shoots through my burning body. Fear on top of fear. Pain on top of pain. The pain is pounding to the beat of my racing heart. The car, which moments before felt like my tomb, now seems like a safe haven. But I don't know where it is—left or right, in front of me or behind. In the time it takes to inhale, I've lost touch with the only thing that was keeping me grounded. My knees weaken and fear engulfs me. I realize how very lost, confused, and helpless I am. My flailing hand finally touches the car again, and I take another breath, this one deep, thanking God that I'm on familiar ground once more.

I imagine an angel touching my shoulders, but then I recognize the feel of Jim's arms and hear his voice. With his right arm around my waist and his left arm holding my hand, Jim rushes as he leads me, blind, across the parking lot to a water faucet.

"Here, baby, get down on your knees. I've got to get you washed off." Without arguing, I do what Jim tells me. Cupping his hands under the faucet, he splashes water all over me, trying to wash away a madman's destruction. I'm screaming in pain again. My clothes are still coming off in strips. My skin is smoldering.

Jim's voice is pure controlled calm at this point. I'd always known what a brave person he was, but now his strength is really coming through.

"Cindi, everything's going to be okay. An ambulance is on the way. Help is coming. Hold on. Just hold on."

The ambulance takes time, but the acid wastes none. It is ravenous and merciless and has no saturation point. It burns through surfaces until it is washed away.

"Am I dying?"

I hear a voice I don't recognize. "No, you're going to be okay. You are going to be just fine. I'm Richard, head of security here." Richard joins Jim in the frantic attempt to rinse as much of the acid off me as possible. After a few seconds of using only their hands, they realize the futility of their efforts.

"Wait a second!" Richard says. "This travel plaza has a truck stop—there are showers. Let's get her inside."

They lead me, wet, half-naked, and in complete shock, to the showers. Each step brings more pain. *God, I need your help. I know you have an army of angels and I need to ride on their wings.* He answers my prayers by bringing me these people. But the prayer I need answered most is for the ambulance to get here.

The water from the showers is not the welcome relief I had hoped for. The changing temperatures, combined with the fact that my top layer of skin is gone, brings its own new agony. First screaming, "It's too hot!" and in the next breath, "It's too cold!" and then pleading for someone, anyone, to get me help, I continually ask everyone around me, "Am I going to die?" Once again, I am sure that this is what death must feel like.

Jim himself is in need of attention. The acid has burned his head, right arm, right shoulder, and stomach. One of the female clerks stays with me while Jim is led to another shower.

Her voice is calming. "Just hang on a few more minutes, sweetie."

"It hurts. It hurts so bad."

"I know. I'm so sorry."

"Where's the ambulance? What's taking so long?"

The clerk explains that it has to come from Blanchard, fifteen miles away.

Some of my clothing is still on me, including my bra, panties, strips of the red Capri pants and chunky white tennis shoes I had bought specifically for my vacation. Only one thing makes it through without harm—those shoes. I actually had resisted buying them, because they seemed too trendy for a grandmother, but my daughter Shelli had insisted they were the perfect complement to my outfit. Getting dressed that morning, I almost put on open-toed sandals, but that hadn't seemed fair to Shelli. I would soon find out how that one small decision would play a major role in this unfolding drama.

In the distance, I can hear the siren. Like someone who struggles for hours to stay afloat in the ocean and relaxes the moment help arrives, only to drown before she can be saved, I slip into unconsciousness.

The clerk rouses me just as a man approaches my shower.

"Is the ambulance here? I thought I heard a siren."

"No, honey, not yet. It's just the police."

"Officer Gary Norman, Newcastle Police Department," he says. He tells the clerk to keep turning me around under the showerhead.

"Make sure to get some in her mouth, too. She probably screamed on impact, and who knows how much she's ingested."

"Am I naked?" I ask, now that I know a man is nearby.

"No, you've been taken care of. You're covered," Officer Norman answers. He sounds calm and collected, but when I met him later I learned that his gut instinct told him he was looking at a dying woman. He also told me that when he'd driven up to the travel plaza and seen my car, he found it hard to believe his eyes: he'd never seen anything like it in all his years of police work. It looked like an enormous rock had smashed the front window, but it was obvious that something hot had melted the glass. Whatever it was had turned the inside of the car black, like a trash fire. After one look at that car, and another at my skin, he knew he had a major emergency on his hands. He immediately suspected foul play.

It sounds like everyone is yelling. Over all the commotion, I can still hear the police questioning Jim in the next shower stall. "Where did this happen?" "What did you see?" "What can you tell us?"

I turn toward the sound of Officer Norman's voice. "Please, please," I scream, "call my daughters!" I'd been protecting them since they were babies, and even in my condition, all I care about is taking care of them. Even though they are grown-up women, in my mind, they're still my little girls.

"I don't think I'm going to make it. Call Angela and have her call Brandi because she's the closest and she can get here first."

"Do you want me to just call Brandi?"

"No, let Angela call her sisters. Whatever you do, don't call Shelli. She can't take this kind of news. I'll give you her number, but don't call it. If Angela doesn't answer, then call Brandi next. Whatever you do, don't call Shelli."

I'm so busy screaming at Officer Norman, I don't hear the ambulance arrive. One EMT checks on Jim, while the other rushes to my side. I'm vaguely aware that I'm being lifted onto a gurney and keenly aware of the piercing wind as I'm rushed to the back of the ambulance.

The paramedics ask me where I want to go.

Dazed, I reply, "The closest hospital you can get to."

"What about the burn center at Baptist?" they suggest.

"I really don't care where we go. Just get me help."

I hear radio chatter between the hospital and the paramedics.

"We have a female patient, extensive chemical burn, lots of pain."

Lying there, all I want is to wake up from this nightmare. *This can't be happening to me.* This kind of thing only happens in the movies. But, I realize, it *is* happening to me.

Jim climbs into the ambulance beside me. "Hang on, Cindi, you're going to be okay." I'm not convinced.

While one medic jumps into the front seat, the other works on removing what's left of my clothes. My bra straps are so soaked with acid, they're eating the flesh on my shoulders. The skin that has melted around them makes removal very difficult for the medics and unbearably painful for me. They pull at each piece. I beg them to stop, but my pleas fall on deaf ears.

"I'm sorry, but we have to get this off." They cut the bra, inch by inch, until it's finally gone.

"Are we clear to give her morphine?"

The medic looks for a place to put the IV and can't see one. I have burns all over my body—my eyebrows, eyelids, nose, lips, cheeks, chin, neck, chest, breasts, stomach, arms, hands, fingers, and the right and left legs from the knees down. Amazingly—I believe miraculously—the skin on my thighs was unscathed, although we didn't realize it until later. And later, having this unburned skin would prove crucial to my treatment and recovery. The burns are so bad, he can't see a vein where he'd typically insert the needle. He also knows that all it takes is a tiny point of entry into burned skin to cause a deadly infection.

I hear him say, "Let me check her feet."

Jim and the medic unlace my shoes, the ones I almost didn't wear. Somehow, while everything else on my body is soaking up acid, the chunky white canvas tennis shoes do not. Was it divine intervention or just a lucky choice that I wore them instead of a pair of sandals that would've left my skin exposed? Those shoes not only protected my feet from getting burned, they were the only reason I would eventually get a reprieve from the pain. But first, I scream out loud as the medic pierces my toe with a needle. I hate needles, but I had no idea that an IV in my foot would be this painful. For that instant, I have no doubt that I am alive.

"Are we there yet?" I am desperate to get to the hospital as quickly as possible because I'm convinced that's the only place that can save me.

The ambulance starts to roll and then stops.

"We haven't even left yet," Jim answers. Someone is talking to the paramedics, and I can't believe what I'm hearing.

"Don't leave. We've called HAZMAT. We have to wait on them."

Oh my God. This can't be. Hazardous material. We have to wait on guys in white puffy suits and gas masks, and all I want is for this ambulance to move.

I start to feel the effects of the morphine. My toes go numb, then my legs, then my arms. For the first time in almost two hours, my screams subside and my body relaxes. I feel like I'm floating, but the ambulance isn't moving.

MY MIND DRIFTS BACK to when we were planning this vacation to California. Jim, who had convinced me San Diego was spectacular, also tried to persuade me to spend the night in an Oklahoma City hotel near the airport. He suggested leaving on the evening of June 4, but I wanted to head out first thing June 5. My reason for the later start was to be home in Duncan for something important on Monday night. My four-year-old granddaughter, Kennedy, was going to play in her very first T-ball game, a child's version of softball. Kennedy is my first grandchild, the light of my life, and I didn't want to miss any of her games.

In the end, we decided to stay in Duncan and use the hotel money on something fun in California. I also promised to stay awake and keep Jim company during the drive to the airport. After we made the decision to leave in the morning, we never looked back.

The game was great. Kennedy made a few base hits, and I was glad I was there to see them. I'll never forget the walk back from the ball field back to the parking lot—Angela and her husband, Steven, who was holding their newborn son, Carson; Kennedy, who was running ahead with all her spunky, funny teammates. The short journey was filled with jokes, laughter, and a feeling that all was right with the world. I had no idea that in just a few hours, these would be my last memories of a normal life.

So early Tuesday morning, and I mean early—2:00 A.M.—the alarm clock sounded. I jumped out of bed with excitement. As a single mom, I hadn't treated myself to many getaways. I worked too hard, tried to save as much money as I could for the girls, and, before Jim, did not have a partner to travel with.

I hadn't slept very soundly and couldn't wait to enjoy a week of relaxation in the sun. We had so many reasons to go to San Diego— the zoo, the beach, and the Padres. We didn't care much about the "Pads" per se, but being big Braves fans, we'd hoped by some miracle that our team would be taking on the home team the week we intended to be there.

Our flight to San Diego from Will Rogers World Airport in Oklahoma City would take off at 5:30 A.M., with or without us. Jim and I wanted to make sure we got to the airport in plenty of time, which meant leaving home by 2:30. He packed the trunk of my black Pontiac Sunfire with way more stuff than we actually needed. I took a look at my cute little car; it was the first new car I'd ever owned, and I was proud of it.

Jim slipped into the driver's seat, I sat beside him, and we left the town of Duncan sleeping behind us. We were right on time, but about to find out how disastrous our timing truly was.

The breeze coming through the half-lowered window, the one that was supposed to keep me awake while Jim drove, was relaxing. It was tough not to drift off to sleep on a dark, straight, and empty highway. But I had promised Jim I'd stay awake.

"I don't ever remember the night being so black," Jim struck up conversation to keep us both alert. "There aren't any stars out. I don't even see the moon."

"Jim, I know I'm not supposed to, but really, wake me if I go to sleep."

"Don't worry, Cindi. Your snoring is enough to keep me as alert as I need to be."

Even though I swore to Jim I wouldn't, and although he knows I always do, I fell asleep daydreaming about the part of the Pacific I would claim for the next five days.

Jim glanced at the upcoming overpass. Our car lights were the only illumination anywhere, but out of the corner of his eye, he saw a flicker. We were halfway to the airport, but halfway was as far as we'd get.

IN THE END, the ambulance doesn't wait for HAZMAT, but they and the fire department are waiting for us when we pull up to the emergency room at the Baptist Burn Center. The ambulance stops with a jolt. *Thank God we're finally here.* There are only a few feet between the back of this ambulance and the front doors of the ER, but something's wrong. For some reason, nobody's moving.

They're not going to let us in. It feels like I should've been rushed into the burn center hours ago, but once again, I'm left to wait. *Does anyone realize what I'm going through?*

Jim does.

Still wrapped in only a sheet, he jumps out yelling, "What is the holdup? Get her inside." For a Scotsman with a quick temper, he actually tones it down, not saying what he'd really like to, hoping to improve our chances of getting help quickly.

A firefighter answers, "They can't let her in until they know what burned her."

"I can tell you it was acid. Get my clothes out of the bag. She's showered. She doesn't have anything chemical on her. Just get her inside," Jim shouts. He'd had the presence of mind to save his own clothes in a plastic bag after taking a shower at the travel plaza, thinking that authorities could use them to test the chemicals later.

"Let her in. She needs help. I'm telling you, it was acid." His is the voice of a confident and deeply concerned man who doesn't need a scientist to tell him what the substance is and what I need.

"How are you so sure?"

Patiently, Jim explains that in the seconds after the attack, a small drop had landed on his lip. Without thinking, he licked his lips, tasting the acrid liquid. It brought back memories of being a kid working on car batteries and accidentally rubbing his dirty hands on his mouth.

"I'm telling you, it's acid."

Inside the ambulance, I'm crying, "Please, please somebody help me."

Finally, by the grace of God, after more than two hours of suffering, the hospital lets me in.

⁓

JUST AS I'M wheeled toward the emergency room doors, a plane with two empty seats flies overhead.

Our five-day vacation will turn into an eighteen-day stay in the burn center. What I thought was the start of a relaxing getaway was instead the beginning of a journey beyond mere survival.

As the automatic doors to the ER swing outward for us to enter, I feel a blast of freezing cold air. *God that hurts.* I grab my face. The air-conditioning would normally be welcome on a summer morning, but this time it brings on a new wave of pain. I don't know if I'm saying anything coherent, but I can hear everyone else, and it seems that people are everywhere. A nurse puts a fresh wet cloth on my face.

"Please, am I dying?" I whisper from under the cloth. Or maybe I scream. This voice steeped in pain is not my own. "Am I going to die?"

"No, honey, you're not going to die."

I kind of believe her. Then I wonder if she's just saying that. I ask someone else.

"You've been burned. It's a chemical burn."

"Is it going to burn my organs?" I can't imagine anyone could survive that.

"Look, you're in the best hospital in the state, and we're going to take care of you."

Her words should be comforting, but I'm consumed with worry. I always thought this hospital cared for people with burns from fires, not from acid. They barely let me in the front door. How can they possibly know what to do with an acid-burn victim? How many of their patients get acid tossed on them as they're driving down the highway? Now I'm really in trouble. In an instant, I feel like I'm back at the travel plaza in the dark, silent night feeling my way toward nothing. I'm slipping away again. I'm alone again.

"Where's Jim?"

"He's okay. He's doing much better than you are."

They leave me alone. No one is talking to me now. I don't know where I am, what's around me, who's around me. I wish they would talk to me, walk me through everything they're doing.

Again I ask myself if I'm dying. I feel like, if I'm lying here not dying, then people should be all over me trying to save me. Maybe I'm unconscious. Maybe I'm dead.

I hear rustling near my bed, presume it's a nurse, and ask my most pressing question:

"Are my daughters coming?"

"They've been called."

The firefighters return to their posts, the ambulance waits for its next call. For them, this crisis is over, but for us, it's just beginning. True to Officer Norman's fears, I have inhaled acid.

Will the girls get here in time? Will I live to see them once more?

2

Gathering My Army

TIME IS STANDING still for me; I feel completely powerless. All decisions about me are being made by total strangers. I don't have the ability to call my daughters myself, I don't know where Jim is, and I can't even see what's happening to my own body. Having taken care of myself for so long, it's difficult to accept that I have no control. There's no one here to help me. I am alone.

Even though I feel abandoned now, once my girls and I are reunited, I find out that forces were mobilizing. Officer Norman did exactly what I'd asked—he called Angela first. When the phone rang so early in the morning, Angela's husband, Steven, was in such a daze that he feared my plane had crashed. Officer Norman explained I'd been burned in an accident but didn't say much more. I'm glad he downplayed the situation because I don't know how my children would've made it to the hospital in one piece if they'd known the truth. He said not to rush, that when he'd last seen me I was talking to the ambulance driver. Even so, Angela trembled as she spoke with him, almost dropping the phone. She said she tried not to cry but couldn't stop the tears.

With my army: Kendra, Angela, me, Brandi, Shelli, and Kennedy

Angela didn't need a police officer to tell her what to do next. She called Brandi, asking her to hurry to the burn center, because she did not want me to be alone. They would've wept buckets if they'd known just how alone I was.

Brandi immediately jumped out of bed, called her best friend, Kendra, who is like a fourth daughter to me, and waited on the curb in the darkness for Kendra to pick her up. Brandi tells me that she thought I had leaned up against something hot, burned my hand, and missed the plane. During the car ride to the hospital, Brandi prayed that's all it was.

After they arrived at the ER, Brandi says they spotted an open door at the end of the hallway. Inside the room, she could see Jim sitting up in bed. He appeared to be fine. She couldn't see the extent of his injuries.

Before they could walk into his room, Jim pointed his thumb to the right and said, "Hey, kiddos. Your mom's in there."

Thinking that Jim was okay, Brandi tells me, the weight of her worry lifted. That's when Brandi walks into my room expecting to see the same.

I hear a whimper.

A small gasp.

I know that sound.

Before she ever utters a word, I say, "Beebop, I'm okay. I'm okay. I'm okay."

My baby cries. "Oh, Mom" are the only words she can get out. Brandi, my bubbly, bouncy, youngest daughter, so full of life and never at a loss for words, is suddenly speechless. I can hear her trying to stifle her sobs. She doesn't believe me. She knows I'm not okay. She can see what I can't, that I'm lying there, my body mostly covered by a sheet, like a corpse. My face is red and blistered, but my skin doesn't look burned; it looks melted.

Brandi doesn't know that I can hear her whisper to Kendra, "Is that life support I hear beeping?"

Kendra says, "I have no idea. I was hoping we'd walked into the wrong room."

"Oh my God," Brandi sobs, "I can't live without my mom."

Normally I would reach up and put my arms around her and tell her I'm going to be okay. But I can't move my arms. She'll just have to believe my words, although I'm not totally convinced myself.

I ask her, "Will you check on Jim for me?" I'm worried about him. Even though they've just seen Jim, Brandi asks Kendra to do it to satisfy me. It's overwhelming for all of us. Brandi won't leave my side.

"Mom, where can I hug you?"

We are a family of huggers. I truly believe there is healing power in the human touch. It is a gift we give ourselves and our family and friends. Even though it seems impossible because of my burns, I want to feel my daughter's energy. Instinctively, Brandi wants to give it to me, but there is no obvious place to hold on.

"Oh, Beebop, no, don't touch me. I don't want you to get burned." Since the shower at the travel plaza is the only scrub-down I've had, and no one in the ER has washed me off, I fear that the acid is still on my skin.

"Why would someone do this to you, Mom?"

"I don't know, baby. But I do know he ruined those cute Capri pants I just bought."

My feeble attempt at humor is meant to calm Brandi, but she's feeling too sorry for me to laugh. Whether I sense her fear and try to lighten the moment or it's the morphine, I try again.

"I'm just wondering if I have eyebrows."

"Mom, we've got bigger fish to fry than that," she says, glancing down at me.

After a moment, she says, "Mom, you know how much I love you, right?"

"Without a doubt, Brandi, I know how much you love me."

Brandi reaches over to rub the top of my head. Normally I hate to have my hair stroked. A lot of people find it comforting. I find it annoying. But not this time. It feels good to have her touch me.

"Mom, I can see the palm of your hand, and I don't think it's burned."

Brandi slides her hand under mine. Feeling secure, I drift off to sleep.

I wake to the sound of voices. I hear nurses tell Brandi and Kendra that they're taking me upstairs, which brings on a whole new wave of fear—fear of the unknown.

As I'm wheeled down the hallway, I scream.

"The air! The air! Oh God, it hurts!"

I wish I could hold back for Brandi's sake. I know hearing her mother cry is overwhelming. But it feels like freezer burn, like another splash of acid. I gasp. It's killing me.

A nurse takes the sheet covering my body and pulls it up over my face. The searing pain subsides for a moment. All the while, Brandi has not let go of my hand.

When we reach the third floor, the nurses tell Brandi and Kendra they can wait in the room reserved for me while I go to what's called the tank room for the first of countless trips. Brandi and I don't want to be separated from each other, but it sounds like a medical necessity.

"Mom, we'll see you as soon as you come out."

The mother in me says out loud, "When you call your sisters, tell them I'm fine. Really." But Brandi can't be convinced. This is her worst nightmare coming true. Two days ago, Brandi dreamed that something terrible happened to me. Whatever it was, she had a vision that she and her sisters were in a hospital crying over me. When Brandi woke up, she called me in tears. Later that day, I e-mailed her a card from work that read, "Just wanted to let you know I'm alive and well and I love you."

Reluctantly, she watches the nurses wheel me away.

Knowing Brandi, she will reach her sisters on their cell phones, but she won't do as I've asked. She'll tell them exactly what she's seen, and that terrifies me.

STILL UNABLE TO SEE, I feel myself being dipped into a huge tank of water, and as gently as the nurses can, they scrub down my entire body. The pain is unbearable. The wind was torture enough. Even with an extra dose of morphine, this is worse. I cry out, begging them to stop. But they have to keep going. They have to wash off every drop of acid, because it keeps burning until it's entirely washed away. While the third-degree burns look the worst, the second-degree burns have exposed nerve endings, and there's no drug in the world that can kill that pain entirely.

Once the acid is washed off, it becomes a matter of watching and waiting. The nurses explain that because these are chemical burns, instead of fire burns, it can take up to three days for the burns to "declare" themselves—how big and deep they will be. In the meantime, all the nurses can do is keep them clean and covered in layers of ointment and gauze, and try to get my pain under control with a constant morphine drip. My body is swollen, especially my face, which is why I can't open my eyes. I still don't know if the acid has blinded me permanently.

I'm taken to my room, where Brandi and Kendra are waiting, and lifted onto my bed, and immediately a nurse begins an assault on my blind eyes. She literally shoots saline directly onto my eyeballs. It is beyond uncomfortable. Four gallons in two eyes. I cry throughout the eyewash. The girls are helping the nurse by holding my eyelids open. They are crying, too.

I scream, "Stop! Please! I can't take anymore."

And just as I've reached my limit, the nurse says, "Okay, it's over."

Out of the slits in my eyes, I barely make out a face just above mine. It's Brandi's.

"I can see you, Beebop. I love you."

"I love you, too, Mom."

"Where are your sisters?"

"They're on their way, Mom. They're on their way."

The phone next to my bed rings. Kendra answers and puts it up to my ear.

"It's Robin," she says.

Hearing my own sister's voice is comforting. Even though she's my youngest sister, I've always relied on her strength and wisdom. She knows just what to say.

"Hang on, Cindi. I'm coming. You're going to be okay. Don't worry about anything."

"It hurts so much, Robin." I know it must be tough for her to hear this, being so far away in Dallas, but I feel so swallowed up by pain I can't think of anything else to say. By now, we're all in tears—me, Robin, and Brandi, but especially Shelli, who has just walked in with her husband, Chad, and heard our conversation. Her eyes are bloodshot, and her face is taut with gut-wrenching concern.

"ShuBug, please don't cry. I need you to take care of my grand-baby."

This, of course, has the effect of making Shelli cry more. She's newly pregnant, and I worry I won't live to know her first child.

"I'm sorry, Mom." Shelli reaches over to stroke my hair. "I'm so proud of you."

"I'm so glad you're here. I love you so much."

We're desperate to hold each other, but the sight of me wrapped head to toe in gauze leaves Shelli feeling there just aren't many places she can touch me without hurting me.

Instead, she holds up my wallet. "I got this out of your car." I can see her fighting to control her emotions.

"How did you get that?"

"I stopped to use the bathroom at the same travel plaza. I saw a car wrapped in police tape, and I realized it was yours. A detective let me have it."

I can't even imagine what my tenderhearted daughter went through when she stumbled upon the scene. If I could have kept her from any place on earth, it would've been there. She later tells me that the windshield on the passenger's side, where I'd been sitting, had shattered inward. Shelli had fixated on that web of broken glass and called Angela immediately to let her know what she'd seen. Unlike Brandi, Shelli arrived at the hospital already knowing the situation was serious.

"Is Angela here yet?" If I'm not meant to live, I need to see all my daughters before I die.

"I just talked to her. She's on her way—it took her a while to call everybody."

At last, I hear Angela's voice asking Shelli, who's standing near the door, "Is she okay? Steven and I saw the car."

As she steps into the room, she instantly tears up. She's shocked at what she sees. As an experienced nurse, she can usually remove herself from the situation. But this time is different.

"Oh my God, Mom. I would've gotten here sooner if I'd known how bad it was. When the police officer called, he told us there was no reason to rush. I'm so sorry."

"It's okay." At this point, I'm crying with relief. All three of my girls have rushed to my side. My army has gathered. I won't have to fight alone in the battles to come.

It's breaking my heart that my kids are so sad. I have to let them know that I will do anything in my power to be there for them. I think about each of them coming to the hospital with so much uncertainty, and it takes me back nearly two decades to my own tearful trip to a Wichita Falls hospital. My mom was hospitalized and I didn't know why. I cried the whole way. I had no idea a heart attack would claim her life at fifty-nine. She would be gone before I could see her one more time. So I know what my girls went through as they drove to the ER. I thank God for this moment right now, that all of us are here together.

I moan.

"Mom, what's the matter?" Angela asks.

"My breasts hurt."

"What? Where?" one of the girls asks. They pull back my gown, but don't see anything unusual on the surface of my breasts. Bandages cover most of my chest.

"Under. It's underneath them."

Through the morphine, I feel sharp stinging under both breasts. When a nurse lifts them up, she finds shattered glass from my windshield embedded in my skin. She and the girls pick out some of the larger pieces but have to leave hundreds of tiny shards to work their way out naturally.

I look up at the nurse and ask, "Where is Jim? When can I see Jim?"

I haven't seen him since we were brought into the ER, and I am desperate to talk to him in case I don't make it.

She tells me he's being settled into a room directly across from mine.

A few minutes later, Jim walks in.

In a tender voice he asks, "You okay, baby?"

"Yeah, but are you?"

I'm thankful that his injuries are not quite as severe as mine, but I know he is suffering. His right arm took the brunt of the burns, along with scattered places on his body. None of this seems fair for a man who had already fought and lost a heartbreaking battle.

I'd known Jim for seven years, having met him right after his wife of thirty years died of breast cancer. They fought the disease for nearly a decade, until it finally took her. When Jim and I found each other, we were both lonely and needed to share our lives with someone. But I'm afraid I'm about to be the second woman he loves whom he will lose. I want him to know that if I die, there was nothing more he could've done.

"Jim, you're a hero. I know you're the reason I'm still alive."

"I just did what I had to do, Cindi. You just do what you need to do to get better," Jim says as a nurse takes him back to his room.

Apparently a lot of people are worried they'll lose me, because my hospital room begins to fill up. It's chaos. I can't make out every face because my vision is still blurry, but I know that besides my girls and Kendra, my best friend, Sharon Smith, is here. At least three detectives from different law enforcement agencies, a PR representative from the hospital named Damon, and Darrel and Deron, close colleagues from work, have arrived to check on me. Like my girls, they all knew that I had been in an accident, but nobody expected what they saw.

Through the haze I can see a room full of heartbreak. Their voices are a sad chorus of tears and whispers and strain. I focus on one figure in the crowd—the regional manager at Cable One where I work.

"Is that you, Brian?" I ask.

Barely able to answer, his voice cracks. "Yeah, Cindi, it's me. You okay?"

"Thanks for coming."

I expect my family and friends to be here, but hearing the emotion in Brian's voice, seeing the shock on his face, and knowing he drove four hours from his home in Sherman, Texas, to be here, makes me wonder if everybody isn't coming to tell me good-bye.

"Am I going to be okay?"

That question prompts Angela to action. She goes to the nurses' desk to demand some answers.

"Can someone please tell us what's going on? Has a doctor been called? Is my mom going to die?"

Angela comes back to my room and calls Brandi and Shelli over to my bed.

"This is what they told me. Mom has a burn index of sixty-nine, which means she has a sixty-nine percent chance of not making it." But she quickly adds, "I know that sounds high, but they say she will pull through. She inhaled some acid, but not enough to put her in the ICU. We won't know for a few days how bad these burns are. One of the best surgeons is coming to check on her tomorrow."

That's all I need to hear. There is hope. Lying in a hospital bed, wrapped like a mummy, I'm making a crucial decision in my recovery. I'm going to push thoughts of dying out of my mind and think only about living. I know I have something to live for, and if I can help it at all, I will live for my children. Even though fear, pain, and uncertainty made me feel like I didn't have a choice, I realize now that

I do. I now choose to believe I will live. I hope if my story teaches anything, it's how vital the choices we make are. Whether we have a few minutes left or a lifetime to live, choose to make that time count.

My daughters inspire me to make this choice. When Jim and I were first attacked, and I could only picture their faces in my mind, I thought so much about dying. But now that they're here and they're real, I see that their love, compassion, and loyalty will bring me through this awful chapter in my life. Some people lean on their spouses, some look to their parents, some rely on best friends, but for me it's these three remarkable, courageous, compassionate young women who have rushed to my side. They are my powerful little army in this battle to survive. They changed my thoughts from dying to living. It might have been just a glimpse that first day, but it would become an inspiration every day in the unthinkable pain of the burn unit, during the struggle of physical therapy, and throughout the dread of more than one dozen reconstructive surgeries. It's not just about being alive; it's about living life.

3

I Am Not a Victim

My journey of a thousand miles began with one small step, then another, and then another. The steps I took were all choices anyone could make. The first steps were to choose to fight for my life in the moments after the attack, and later in the hospital, choosing to believe I would live. Looking back, I can see three crossroads where the paths I took led me to where I am today.

First, I decided what kind of survivor I would choose to be.

Second, I made the choice to heal from the inside out.

And finally, I chose what to do with the rest of my life.

I believe the path to healing is lined with choices. I did not choose to have a gallon glass jar of acid break over me, and only God could choose whether I would live or die. But it was my choice whether my spirit would live or die. As you travel this journey with me, I hope you can take something from my story that helps you when you reach a critical crossroads in your life.

Jim and me, a few months before the attack

WHAT KIND OF survivor would I choose to be?

I decided early on that I did not want to be a victim, that bitterness was not the legacy I wanted to leave for my grandchildren, and that vengeance was not mine.

For the nearly three weeks I was in the burn unit, I couldn't take a bite of food by myself, go to the bathroom on my own, or take a single step without someone's help. It would've been easier to retreat into my own world where I could complain and be resentful toward the nurses and doctors for the painful procedures I was going through. I could've lashed out at God for creating the human being who did this, someone who had so much hate in his heart.

Police reports, medical reports, and news reports all called me a victim. Of course, technically, I now know I was the victim of a hideous crime. For whatever reason, the perpetrator chose Jim and me, two innocent people. From his point of view, he timed it just right. He dropped the jar sixteen feet and hit the bull's-eye.

At that very instant, I became a statistic. I believed I would die from a random act of violence. I would no longer exist because of the distorted whim of a passing stranger. But I survived, and so to me the victimization ended there. It had to. I'd rather be known as a survivor who made something good come from something terrible.

IT'S EXHAUSTING TRYING to wake from this nightmare.

Detective Dale Sparks with the Oklahoma State Bureau of Investigation makes his way through the throngs of people who fill my hospital room, introduces himself, and asks if I can answer a few questions. I want to answer them, but I'm wiped out from the morning's events. *Can this really be happening to me?* It is so surreal to think that I've gone, in an instant, from a tourist on vacation to the subject of an investigation.

"Did anyone know the route you were taking this morning?"

I realize I haven't grasped the severity of the situation until now. I want to scream at him, *Oh my God, no, I don't live in that kind of world! What are you saying, that this wasn't just an accident?*

"Cindi, I need to know, do you have any enemies?"

This question seems even more incredible to me. Even though I'm fixated on Detective Sparks, I'm aware of all the friends still in my room. I think, *People with this many friends around within hours of an accident don't have enemies.*

My voice cracks and gets a little higher as I try to answer him without falling to pieces.

"No, no one knew my route. No, I don't have enemies. You honestly think someone was lying in wait for me?"

"I have to check out every single possibility. I would think you'd want me to consider everything," Detective Sparks says.

I am stunned by his line of questioning. I'd hoped that a jar of acid had fallen off a truck and landed on us by accident. I know he's doing his job, but I'm convinced that there is no way this act could've been anything but random—whether it fell off a truck or slipped through someone's hands. I don't dare say it aloud, since he's the professional, but I can't help but think, *You're barking up the wrong tree. If you're going to solve this crime, you need to take your questions in a different direction.*

As if he were reading my mind, that is exactly what he does. But the direction is not the one I expect. It's even worse.

"What about your ex-husband?"

That question snaps my girls to attention. Even though they've been listening to every word, they now look like someone's punched them all in the stomach.

"No way! My ex-husband wouldn't dream of doing anything like this! Don't even go there. I told you, I don't know those kinds of people and neither does he."

Detective Sparks must realize that I can't handle any more questions today, and he has the answers he needs to get started. He has his job cut out for him. With no eyewitnesses and very little physical evidence, it will take every bit of practical knowledge he has. Either that or a very lucky break. This will not be his only visit to my hospital room, but it's a good thing he's talking to me now, because once treatment for my burns begins, who did it and why will be the furthest things from my mind.

"Before I go, where can I find the other victim, your friend, Jim Maxwell?"

I answer him, but all I can think about is that word: *victim.* I hate the memories it conjures up in my mind. Years ago, I, too, referred to myself as a victim. I played the role in every sense of the word. I became the best victim of all. Detective Sparks may refer to me as a victim as a formality, but I won't see myself that way because that's not the kind of survivor I want to be. It's a lesson I've learned the hard way.

～◦

NINETEEN EIGHTY-EIGHT WAS THE YEAR my husband, David, left me and the girls. I had no job, no money, no education, no parents, and four mouths to feed—not to mention providing clothes and shelter. The girls were seventeen, twelve, and ten. They were heartbroken over the breakup of David's and my marriage. For their sake, I always made sure they came home to a happy face. But it was a different story when the girls were in school and I could have a pity party for myself. Most of the time I'd sob my woe-is-me song in the shower. Something about the way the hot water poured over my face and body opened up the tear faucet. Even out of the shower, I'd cry all afternoon long thinking about my impossible burdens, like, *How am I going to raise these kids? How will I afford to keep this house? How will I do it without him?*

Having lost my dad two months earlier, and my mom four years before that, I felt so lost. I had nowhere to turn and no one to turn to. Everything secure in my life was gone. I had let someone else make decisions for me for so long. Now it seemed easier to hide away, feeling sorry for myself, than to wake up each morning and face the world. It was easier to be a victim.

Where had it all gone wrong? I thought my marriage had been a good one. For eighteen years, I never really felt unhappy.

Except that I hated that ash paneling.

Our living room was lined with ash paneling. It was the first room you'd see when you walked into our house, and it was windowless. The ash paneling made it lifeless. It just made me sick.

"We can't take that down," David would say. "That's good ash paneling. We're not going to waste it."

"But it's old and ugly and out of style," I'd argue.

"Oh no, it's still good."

I'd wait a few months and try a different approach.

"What if we take it down and store it in the garage? That way we won't actually waste it." My reasoning didn't make sense to him.

"Don't be ridiculous," he'd say. "That's good ash paneling."

"How about we leave up THE GOOD ASH PANELING and just paint it a different color?" No change. I was forced to live with it for years.

The ash paneling proved that, in my marriage, my feelings didn't matter. The more I was forced to live with the stuff, the more I grew to detest it. I couldn't change it, wasn't *allowed* to change it, and so the "good ash paneling" became a sore spot with me.

One afternoon after David left, I was in the middle of a good ole pity party. This time, instead of in the shower, I was boo-hooing hard on the couch in the living room. The sobbing came right from the pit of my stomach.

"Life is not fair. I hate David. I hate not having money. I hate that my kids are suffering.

"And I hate . . . this damn ash paneling!"

Suddenly, through the flood of tears I looked around and saw my reality very clearly. I had become somebody whom I didn't like anymore. I had ceded my power years before, and I was as much to blame as anyone. And now was the time to reclaim it.

Just as suddenly as the realization came, the crying stopped.

"Well, you know what, he's not here anymore, and by God, it's coming off!"

In the divorce, I had retained custody of three girls and two tools: a hammer and a crowbar. I went to the garage, dug them out, returned to the house, and popped the front door off. I decided the baseboards should come first, then the trim pieces, and finally that paneling! With each piece my resolve grew.

"Here's to you!" I shouted out loud. "Here's to your new life! Here's to my new life! Here's to doing it my way for a change!"

Pretty soon, the biggest mess of paneling you've ever seen had accumulated outside my front door. Old bent nails, curlicues of glue, and wood chips fell on me like confetti. It was New Year's Eve in March.

"Here's to a brighter room! Here's to my way! Here's to my life!"

By this time I was laughing hysterically. All that so-called good ash paneling never looked as good in my house as it did in a scrap pile in my front yard. Inside, I had the ugliest walls you'd ever seen. Dried glue, no texturing. But inside of *me,* I'd never felt more free. I'd set a goal at lunch and accomplished it before dinner. I never knew home improvement could reflect self-improvement. That a housecleaning would amount to a soul-cleansing.

Shelli and Brandi came home from school just as I was in the middle of it all. I don't think either of them blinked. Brandi turned to her wide-eyed and speechless sister and said, "I guess Mom is feeling a lot better."

I *was* feeling better. That day led to an epiphany that woke me up. I had become such a victim. It was everybody else's fault but mine. I was blaming people and circumstances for everything wrong in my life. I started to wonder how my girls would learn anything that would help them in life if I couldn't help myself. Yes, things happened that were beyond my control. But that day it hit me like a ton of bricks. I am not helpless. If I don't like the me I see in the mirror, that is something I *can* control.

⁓

Taking a look at the me in the mirror doesn't even occur to me while I'm lying in this hospital bed. I'm certain my girls don't want me to see how terrible I look. I'm in too much pain anyway to lift my arm to hold a mirror, and my vision is still so blurry, I probably wouldn't recognize my face. Even my own sisters have trouble believing it's me.

At their first sight of me, my sisters, Karin, Jamie, and Robin, gasp out loud.

It is the same reaction, every time. Disbelief. Shock. And always that gasp. Everyone comes not knowing what to expect, because the information is so scarce at first. Everyone who visits feels some relief until they see me. Yes, I'm alive, but now they wonder, for how long?

Outwardly, they try to hold themselves together. But it's too much for Karin. She's crying next to my bed.

It takes everything Jamie has not to do the same. Under her breath, she keeps repeating, "Oh my God."

Robin holds her tears back, too, trying to be strong for me.

None of them can get over the size of my head and the color of my face.

Tears form in the corners of my eyes. "Oh, it hurts so much."

Each of them searches for just the right words to say.

"You're going to be fine, Cindi," Jamie says. I think she says it to convince herself as much as anything.

"We are here for you and we love you," says Robin. "You have amazing strength, Cindi."

Seeing me in so much pain is too much for my sisters. Just when they think they can't take any more, a nurse walks in to check on me. Karin and Jamie retreat to the hallway. Later I find out that they warned their husbands that my face was unrecognizable. And then they burst into tears.

Seeing even some of what this is doing to my family and friends, I sense that I have to do everything in my power not to sink into the helpless, powerless, unlikable person I'd been years before. I don't have to see myself today to know what I learned from all my yesterdays—that my family has always counted on me to be strong. I can see it in their eyes; they need me to be okay. I've been motivated by that need before.

~⌐

My girls needed their mother to be strong, and soon after the ash paneling came down, I began to see myself as industrious. Since I didn't have the money, I figured out how to do the tough jobs around the house by myself. I would decide first what had to be done, and then what it was going to take to do it. I would read and study and ask questions by the thousands. It would've been so much easier for my ex-father-in-law to just do the things himself, but at my request, he spent hours giving me instructions and allowing me to make my own mistakes. Preserving a good relationship with my ex-in-laws was difficult for all of us at times, but after being part of each other's lives for twenty years, we both worked hard at it.

So, with his guidance, I wired lights. I changed an alternator in one of the kid's cars. I took out a wall, cut Sheetrock, and learned how to use a welding torch to fix a leaking shower. I could do anything!

But I still didn't have a real job.

For the first time in my life, I faced the reality that I'd have to make money. David had been paying the major house bills, but as soon as the divorce was final, I'd have to take over. *How will I pay taxes? How will I buy the girls school clothes? How will I keep the lights on at night?*

I considered selling the house. I knew it would be unsettling for the girls, having lost their father and now the only home they'd ever known, but I had no idea how to hold on to it. Not knowing what I was contemplating, Brandi made a simple request one afternoon out of the blue.

"Mom, please don't ever make us move."

It was such a desperate plea from my littlest daughter. I couldn't let her down. So I set another goal: keep the house at all costs. I didn't know how I'd do it. People ask me today how I did it, and I still don't really know. I mowed yards. I took on a paper route. I cleaned houses. I watched other people's children. All after a full eight-hour day at the new job I'd managed to land.

It was enough to get by and help put Angela through her first year of college. Her dad agreed to help pay for some of her college expenses while paying his child support for the other two. He helped out that first semester, but just days before she was to return to school for the second semester, he changed his mind and decided there would be no more money coming from him for college. I had no idea what I was going to do. I cried a thousand tears for the thousands of dollars I couldn't give to my child. Angela worked so hard in her studies, and I wanted to reward her with an education. But even more than that, I prayed that with an education, none of my girls would ever end up in the position I was now in. I wanted for them the peace of knowing that they could help themselves and their children if they were ever forced to. Because Angela was over the age of eighteen, I could not make her father help her. But this was such a dream of mine for them, and I would not take this setback lying down.

"Angela, I don't know what will happen, or how I'll do it, but you will get your college education."

"Mom, when I graduate, I'll get a job and help with Shelli and Brandi. I promise I will."

"No, you won't, sweetheart, but you will go. I'll get you there."

I scrounged every penny I could find and had that girl back in school by the end of the week.

Even though money was tight, and it would've been easy to blame David for our financial woes, one vow I made and kept true to for all these years was how I felt about the girls' father. My only regret about the divorce is that my girls lost contact with their father. I never said anything bad about him to my daughters. I did not want to teach them hate; I wanted to teach them forgiveness. Many times they asked me if I hated him. I always assured them that I did not. I recall one defining moment when I knew that to take the high road, as I had chosen to do, would pay many dividends.

One afternoon, Brandi and I bought a few groceries for a friend who had a newborn baby. Her husband's name was Mike, and he was what is known as an emotional abuser. Mike was there when Brandi and I stopped by. He made some kind of snide remark. And he treated my friend so badly. When we got into the car, I must've said something about what a jerk he was.

"Mom, can I ask you ask something?"

"Sure, Brandi."

"Who do you hate more, Mike or Dad?"

I looked over at my daughter, just eleven years old. She had this look in her eye. I thought, *this is a telling moment in my life.* I was in a lot of pain over the divorce, but I had to rise above it. I did not want to raise my three girls to hate their father. I knew what she needed to hear.

"Beebop, I loved your daddy way too long to ever hate him. So if I had to pick, it would be Mike."

Brandi took a deep breath, which sounded like relief to me. "Oh, okay."

I couldn't know then what I know now, that in teaching my girls forgiveness, I was preparing myself for the truest test.

CONVERSATIONS FROM THE past, like the one with Brandi, are part of the reason the detective's question struck me so deeply. Even though their father and I did not get along after the divorce, there is just no way that he could be involved in this accident. I know the man has no warm feelings for me, but this is way out of the realm of possibility. The more I think about who could have done it, the more I'm convinced it was a stranger who chose us to be his target by chance.

I scan the room, looking at all my friends who have gathered to support me. Even though I don't feel like smiling, it feels good to know I have people in my life who love me like family, since the person I'd love to have sitting with me the most is dead.

I ache for my mother always, but had especially during the days just after the divorce. I still needed her. And then I looked at my children and thought, *I may not have an education past high school, I may not have a job, and I may not have a husband, but I'm still alive, and these three wonderful girls need their mother. They can't continue to live like this.* God must've thought the same thing, because he sent an opportunity that turned our life around. He guided me into fulfilling one of the most crucial goals of my adult life: getting a job.

I am not surprised to see my colleagues Deron and Darrel mulling around my room. They look so sad and so angry all at once. I can tell how appalled they are by what's happened just by the way they pace the room and shake their heads. Some people find that their colleagues are just coworkers and nothing more. But I'm fortunate to work for a great company with great employees who rally around one another. I have no idea if I'll ever get back to my job, but I do know that getting it in the first place was a turning point in my life. As I look over at my sister Jamie, I'm reminded how I got the job that became the catalyst I needed to become fully independent.

~

THE JOB, an advertising sales position in Duncan, was offered first to Jamie, who lived in Enid, a few hours north of there. She told the cable company executives that she wasn't willing to move south, but that she had a sister who lived nearby.

"Is she as good at sales as you are?" they asked.

"She will be," my sister promised.

I wasn't so confident. "Jamie, I can't sell."

"What do you mean? Anybody can."

"But I'm a stay-at-home mom. I don't have any marketable skills. I've had minimum-wage jobs, but nothing that would support a family. I don't think I can do it."

"Well, you better do it because it's a good job, and you've got an interview scheduled for tomorrow."

At this point, I knew my husband wasn't coming back. Our divorce was final. It was all up to me. I'd thought about going to college, but I couldn't see how that would put food on the table now. This was a chance I couldn't turn down. But I was stunned that a job opportunity had come to me so effortlessly.

I interviewed for and landed the job on a Monday, trained on Tuesday, and by the following Monday, I was selling airtime for ads on cable TV. I went from knowing nothing to learning the packaging, benefits, drawbacks, and comparison to other forms of advertising in just one week. It was mind-boggling.

I was scared to death of failure. The job was commission only. Whether I sank or swam, it was all on me, and I didn't fancy myself much of a life preserver. But fear turned out to be a tremendous motivating factor. My first day on the job, I made a 750-dollar deal. My boss hadn't expected me to make a sale for a month or two! Right off the high dive, and I did a backflip. I thought, *I'm going to be okay.* I liked working, getting out, and meeting people. I felt fortunate, relieved, and truly independent for the first time in my adult life.

Just because I had a job didn't mean everything came easily. I lived on a superstretched budget. Then the company filed for bankruptcy. I didn't get a paycheck for a month. No money for a whole month! I paid a visit to a friend of mine, Hal Labyer, a loan officer at the local bank. He had already saved me once right after the divorce when I had no income and no guarantee I could ever repay him. At least I had the semblance of a job for the second loan.

"I will pay you back as soon as I can," I said.

"I know you will. You've proven yourself. Your word is good enough for me, Cindi."

With new ownership, the company survived, and with a new perspective on life, I did, too.

Looking back, I realize the fear I felt about the interview, the training, and the selling was overcome with that first step. I imagined it would be a plunge off the highest cliff onto the hardest pavement, but it ended up being just a step. I don't mean to imply that any of it came quickly or easily, but sometimes we can climb those seemingly insurmountable mountains by setting one goal at a time, even if we have no idea how we'll accomplish that goal. It's not too different from what I'm going through right now. The physical, emotional, and psychological recovery from this crime takes small, sometimes terribly painful, steps. You just have to reach inside of yourself and pull out something you didn't know was there. You find out how much you really have, what you are made of, and who is there to help you.

⌒

THIS IS CLEARLY one of those times. I never knew I'd be tested like this. Taking on a first job was scary, but it was nothing compared to facing down a near-death experience. *Do I have the inner strength to make it out of here? Will taking tiny steps in this case really put me over this mountain?* At this very moment, I can't imagine ever walking again, much less returning to my old life. All I'm focused on now is pushing through the pain, and summoning my courage to not only survive, but to go beyond survival. I have children to guide and grandchildren to love. I am not ready to leave them, and they are not ready to let me go. I know too well what it feels like to be alone in the world.

So many nights, when the kids would spend the night out with friends, I felt immense loneliness. Let me tell you, it's a whole lot easier finding a job than it is finding your way out of loneliness. I cried so much on one of my friend's shoulders that I told her to claim me as a dependent on her taxes and list all those tissues as a write-off. After twenty years of marriage, all our friends were couples. I realized early on that I couldn't and shouldn't rely on my children to fill the void. I encouraged them to go out. Finally, I took my own advice and started dating.

When I knocked on Jim's door two blocks away from my house, I wasn't looking for a romantic relationship; I was just being a friend. I knew that his wife, Joyce, had died a month before. I was one of the first neighbors to be called, and I attended her funeral. When Jim answered the door, he looked terrible. He hadn't shaved, his hair was uncombed, and he had the most forlorn expression on his face.

"I just wanted to check on you and bring you some supper," I said, handing him a plate of roast beef and potatoes.

He apologized for his appearance and said he hadn't left the house. "You're the first person I've seen in three days."

The very next day, on my way to work, I dropped off a project for Jim. Shelli had made a shelf that needed painting. When I handed it to him, Jim said, "Do you know, the one thing I hate most in the world is painting?"

"I don't care. You need something to do."

That night, Jim knocked on *my* door. When I answered it, I saw the Jim I knew. He handed me a beautifully painted shelf.

From that point forward, I found every excuse to help Jim resume living. Any time I'd run errands, whether it was to the grocery store, to the bank, or just to pay the electric bill, I'd stop to pick him up. Anything to get him out of the house, even for just fifteen minutes.

Three months later, Jim summoned the courage to ask me out for a bite to eat. As we walked into the restaurant, Jim, very lightly, probably instinctively, put his hand in the middle of my back to lead me to the table. I felt a zing rush through my entire being.

Oh my gosh, the caring touch of human being. It occurred to me how much I missed the closeness and comfort of a kind, protective touch. That simple gesture released feelings in me that I realized I had shut off. Even though I'm a touchy-feely person with my family and friends, there was a time when I never thought I'd feel *that* way again. *I'm lonely and I'll be lonely the rest of my life.* One touch shattered that fear.

A few nights after the date, I found myself alone at home again, but this time I felt differently about it. I was filled with a kind of peace that was strange but wonderful. In reaching out to help Jim, I found that Jim helped me, too. Trying to sort out what had changed, I wrote this poem.

WHERE DID THE LONELINESS GO?

Where did the loneliness go
that I thought was so safely tucked away?
How did it escape the walls
I built around it that day?
The day I decided to feel no more
so my heart could never break.
The day the pain took control
and engulfed me in this state.
Where did the loneliness go?
It came back out and surrounded me
much to my chagrin
and bombarded me with a fierceness
so my healing could begin.
It sent a friend and his gentle touch
to take its place with me
saying to capture loneliness
means never being free.
He may not stay forever
and hurt may come again.
But now I know where the loneliness went.
It was carried off by a friend.

One touch from a friend did so much for me, and now I am over-whelmed realizing how many friends have come to check on me. I've come so far from that time. I don't worry about being lonely anymore.

But I do wonder if the person who chucked that jar of acid never had a gentle touch. Does he not know the warmth of a loving hand? Is he so sad and cold and filled with loneliness that he doesn't feel pain for someone else? Everyone I know and love imagines this criminal is a despicable human being, a horrible monster. But mostly I believe he must be a lonely, sad, and defeated person. He doesn't care what happens to others because he doesn't care about himself. That's what's so heartbreaking. Not what's happened to me, but the circumstances that would drive another human being to do something so disturbing, destructive, and devastating.

Maybe he sees himself as a victim, one who has been hurt by the world, and all he wants to do is to hurt the world back.

Whether he's a sick stranger or a sad child, I refuse to let him bring my loneliness back. I've been a victim before. Now I'm a survivor. And I don't want to be remembered for giving up on life just as it's handed back to me. I like the reputation I'd earned before I ever knew how my life would change. I overheard Kendra say it once: "You never know when you walk in if Cindi will be watching TV, working on the computer, or ripping out the cabinets with a jackhammer."

By being a survivor, I don't mean that I managed to not die, because we can't control whether we do or don't. But every single person in the world has an inner survival kit. Mine is filled with life experiences that prepared me for the greatest challenge of my life. In it are all the disappointments I've overcome—divorce, joblessness, the prospect of homelessness, lack of education, lack of sense of self, loneliness, sadness, and loss. Without victories over those struggles, I don't think I'd be managing the fallout from a random act of violence—the surgeries, the scars, the stares, and the unsolved crime—as well as I am.

Because of it all, I know one thing—it's not going to help me to ask, "Why me? What have I done to deserve this? Why has God allowed this to happen to me?" Those are not my questions. Although it's natural for victims and survivors of tragedies to ask them, I didn't, and I don't. I don't focus on why. I don't think it's helpful to my recovery; in fact, spending energy wondering about a question that may remain unanswered forever would only hurt me. And to me, it is the epitome of the victim mentality.

It also implies that something so horrible should happen to someone else, that I should somehow be immune to the most difficult challenges in life, but that it is understandable if someone else is not. I prefer asking, "Why *not* me?" Let it be me over any of my family. Let it be me over my neighbors or my friends. *Did someone else deserve it more?* Of course not. It's like when people say the person who did this to me will "get his." "What comes around goes around," people tend to say. If that's true, *what did I do to deserve this round?* No one, and I mean no one, deserves to have misfortune befall them. But as sure as there is life, we will all be tested at some point. It may not be to the degree that I was—it may be more, it may be less—but the crossroads are everywhere. And in facing this, my toughest test yet, I know I'd like myself better in the end if I won, not by being a victim, but by being a fighter. That's the kind of survivor I want to be.

4

Revenge Is Not Mine

TUESDAY, JUNE 5, is by far the longest day of my life. I have made it through the first fifteen hours bouncing back and forth between thoughts of dying to the pain of living. As evening nears and my sisters and the throngs of visitors leave to return to their homes and families, I am exhausted. It's time to finally settle down for the night, surrounded by my girls. I will endure eighteen nights in the burn unit, not one of them spent alone. Angela, Shelli, and Brandi pull up chairs and cots and surround my bed like a fortress. The fear I felt just a few hours ago is eased by the comfort of knowing they are close. Despite all the hell I have been through this day, I'm too tired to think about any of it. Just when I think everyone is asleep, Brandi breaks the silence. "Mom, why would anyone do this to you?"

"I don't know, Beebop. I just don't know." *And I have no idea if we ever will.*

I hear someone whispering. It is Brandi, saying a prayer, asking God to let me live.

As the morning sun comes through the tall and narrow window,

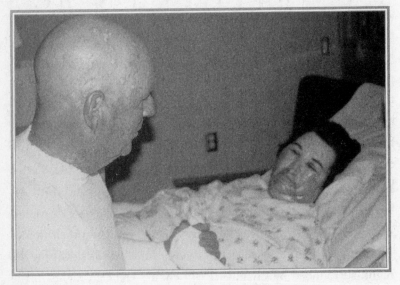

Jim and me, the first or second day

it takes me a moment to remember where I am and why I am here. My surroundings and pain tell me what I do not want to know. This is not just a horrible nightmare that I will wake up from. It is only the second day of a fierce fight to regain control of my life.

I try to force my swollen eyes open, then realize this is as far as they will go this morning. Through thin slits, I see the girls are still asleep. Tears fill my eyes as I think about how much they have been through. Even as they try to rest, the strain of the last twenty-four hours shows on their faces. I know how scared they must be at the prospect of losing me. I have been their strength and sole support for most of their lives, and they are so dependent on me. I know exactly how they feel because I shared the same closeness with my mother. Even though time is supposed to heal everything, I don't think it ever takes away the yearning for your mother when you are hurting. Remembering how important she was to me steels my resolve to make it to the other side of this horror. It's clear that this random act of violence affects more than just two people. My body is battered, but so are the hearts of the people who love me. I have so much more to think about than just myself.

A nurse walks in to check my vitals.

"I'm surprised you're awake so early," she says.

"I wish I could sleep, but I can't. I thought I'd feel better than I did yesterday, but I feel worse," I tell her.

"It looks like some of your second-degree burns have gone third degree. We still won't know for another day or so how burned you really are."

Our voices cause the girls to stir. As they wake, each of them is visibly hit with the same thought I was. *This is a living nightmare. This really did happen.* They see the nurse at my bedside.

"Is everything okay, Mom?" Brandi asks while rubbing her eyes.

"Your mom is fine. The doctor will be coming in shortly to assess these burns."

The swelling has taken over. My lips are turning outward and slowly disappearing into my face. I can hardly move my mouth to talk. My eyelids are curling up as my skin tightens and shrinks. The red, blistered complexion from yesterday is replaced with a bloated, oozing one. I look just as I feel—ten times worse than I had twenty-four hours ago. As the nurses explained, the burns are declaring themselves, some of them more loudly than others. Even though the nurse made this sound normal, I can't suppress the fear that they will continue burning right through to my organs. *When do the burns decide to stop? How do they know they've run out of skin to devour?*

I am in no condition to do anything but lie here. It hurts too much to move. I start to feel nauseated. This, more than anything, gets the girls moving. With all the care they give me, vomit is the one thing they cannot handle without getting sick themselves. Fortunately, Kendra walks into the room to see what she can do to help. It's a good thing because she is the only one with a strong enough stomach to stay with me when I start throwing up.

My breakfast sits on a tray next to my bed, but I can't touch it. I can hear a growing din of conversation outside my room. As the news of our accident spreads, more visitors come to the hospital. My sisters return, as do friends from our hometown and from work. They fill the hallway and take turns coming in to say hello.

Just like yesterday, as each new person gets a first look at me, they are appalled at the sight. Those who'd already spent time with me and wished for the smallest sign of improvement are taken aback. Even Jim, who walks across the hall to visit every few hours, can't believe the dramatic changes taking place, the way my burns are taking over my body. It crushes any hope they have that the worst is over.

Something else is changing, too. The mood in the room starts to turn from shock and disbelief to outright anger. Even though my friends and family know they probably shouldn't say some of these things in front of me, they can't hold their feelings in any longer. They don't even realize what they're saying until after they've said it. Their emotions are just too powerful.

Take Ray, for example. He is the husband of one of my coworkers, and one of the nicest, most soft-spoken guys I've ever known. He comes in expecting to wish me well and raise my spirits, but he's overcome with anger. The moment he looks at me he starts to cry. And then, shaking his head he says, to no one in particular, "Just let me at the son of a bitch. Just let me at him."

My brother-in-law Wayne is furious, too. Normally a gentle man who keeps his opinions to himself, Wayne can't control his emotions. Seeing a woman treated this way insults his manhood. He turns to the other guys in the room and says just what is on all of their minds.

"I'm going to go find who did this and kill him."

It's not just the men. Anger is coming from the women also.

"It's got to be easy to find him in that small town," Jamie chimes in. "I already know what I'd do to him. I could throw acid on him. Not his face, because I don't have that much hate in me, but I could throw it on his lap."

Jamie is just talking, but not Wayne. I worry that if police find my attacker, Wayne might actually gather up his buddies and go shoot the guy. The conversation in the room is all about how to strike back.

Still in his hospital gown with a bandage on his head, Jim slowly moves over to the edge of my bed.

"Is there any way this was just an accident?" I ask him. It sounds like I'm talking through my teeth. My lips aren't moving, but my mind is going a million miles a minute. I know the answer, but a part of me hopes Jim is keeping that possibility open, since no one else in the room seems to be.

"Cindi, there is no one who believes it was anything but intentional. An object doesn't climb up on a bridge and jump off on its own. It had to have some help. Whether it was a prank or a death plot, some yahoo did exactly what he wanted to do. It was a county bridge. You have to know the roads. There is no doubt in my mind that it was intentional." I can hear it in Jim's voice; just like the rest, he's enraged.

But I'm not angry. I'm sad. "That makes me so sick. I can't believe anyone would do anything so awful."

"This a—hole needs to be caught," Jim says. "What in God's name would possess someone to do this? They need to get him off the street.

"You know, I remember when I was a kid and the turnpike had just been built on the other side of our neighborhood. One night after I got home from work, a police officer questioned me about where I'd been, because he noticed the hood of my car was still warm. Why was he suspicious? Some jerk, who happened to be driving a car very similar to mine, was out on a bridge throwing rocks onto cars on the turnpike. I couldn't believe anyone would do such a stupid stunt.

"Here we are now, nearly four decades later, talking about the same crime. Whether it's a childish prank or the work of a sick human being, I don't care. He needs to be caught." All heads are nodding in agreement. Jim's words fire them all up even more.

"He should be strung up," one voice says.

"He ought to be shot," says someone else.

"Just let me have five minutes with him." It's Ray, still shaking his head back and forth.

"No, no, my God, no."

The words come quietly because I'm so weak.

Everyone stops their angry chatter and turns toward my bed. Not knowing if I'm in pain and needing something from them, they stop to listen.

"Don't go do anything. Please don't do that. It's not going to help anything." I can't take all this hostility. I know my friends and family are saying these things to show me support, but I can feel the room heating up with anger, and it's making me feel worse. It hurts to see so many angry faces on people I normally see smiling.

It's even more painful to imagine someone else going through what I am enduring. Even if this person meant to drop the acid, I have to believe he planned to scare us, not kill us. And I can't get past this if I have to worry about my friends going out to hunt down this person. That's violence begetting violence, and I won't support that. I can't heal myself if I have to worry about the terrible things my friends might do on my behalf.

Nobody responds to my plea. They just nod their heads in support, if not agreement, and probably question my sanity at this point. It is difficult for people to understand why I'm not even more angry than they are.

The silence is broken by a nurse who makes her way through the room carrying another flower arrangement. I don't know where she's going to put it. The counters, tables, and even the top of the television are filled with flowers. I love the bold colors and sweet fragrances. They brighten this dull beige room and almost mask the smell of rubbing alcohol. They make me think of the beauty outside, not the ugliness in the world that put me here.

She sets them next to a huge bouquet of yellow roses that had come the day before. The moment I saw those roses, my immediate thought was the Yellow Rose of Texas. *They have to be from Phil.* Jim had read the card aloud.

"You concentrate on getting better. I'll find the bastard who did this. Love, Phil."

I wanted to smile but it hurt too much. I am in no shape for a fight, and it feels good to have my brother-in-law Phil on my team. He has the armor: the ability, the desire, and the means. *If anyone can catch the "bastard," it'll be Phil.*

I begin to realize that there is a distinct difference between going out to catch the person who did this and going out to kill him. I don't mind the idea of him being caught or of people doing everything they can to help law enforcement bring this person to justice. He needs to be stopped so he can't hurt anyone else. But I don't want him to suffer like I am. I don't want him to die. I know I should hate him, but I just can't.

In the corner of my room, Shelli flips on the TV quietly. After a few minutes, there's a story about the acid attack on the news.

"A random attack has Dr. Phil facing his own personal tragedy. All the experience in the world couldn't prepare him for this," reads the anchor. Phil did not yet have his own show, but he had become a favorite on *Oprah,* and a more frequent subject of news reports.

"I want to raise the awareness, because I know there are people out there who know who did this," Phil looks straight at the camera as he talks. "I know there are people in communities who can give us leads, information to get these people off the streets before someone gets killed."

Then his tone softens.

"Cindi is in unbelievable pain. She is a single mother who has put three children through college by working three jobs. She is an absolute model citizen, one who should inspire us all.

"I'm here to try and draw attention to this situation and catch the criminals who have attacked Cindi in such a ridiculous fashion. I've also put up a ten-thousand-dollar reward for their capture," Phil announces.

Shelli's jaw drops. She turns to me and asks, "Mom, did you know there's a ten-thousand-dollar reward to catch the guy who did this?"

"No, I didn't know that. But I'm not surprised," I say. Phil is coming to my rescue when I need him the most. He has always been a great defender of mine. I hadn't even considered how much a substantial reward could help the investigation.

Actually, since yesterday, I have not thought of life outside this hospital. Seeing Phil on TV fighting on my behalf, watching the girls taking control for me, and being surrounded by my sisters and friends put a different perspective on this accident. I am going to have to let people take care of me for a while. It is not easy giving up control. I don't want to cause so much work for everybody. Life has taken such a dramatic turn in two days, not just for Jim and me, but for everyone. I know in my heart we are blameless for the crime, but I almost feel guilty, as if somehow *I* could have prevented all the hurt and anger.

I don't want anyone to fuss over me. But I can tell by the news coverage that the shock of what happened isn't restricted to my hospital room.

From the moment I was brought into the burn unit, my girls had monitored the news. At first they hoped it would explain what had happened on the highway and whether anyone had been caught. My little car with its smashed front window was all over the news, but no one had heard from me or my family about how we were faring.

My friends decide to give the girls some time alone with me. Just as they clear the room, I hear a knock on the door and see Damon walk in.

"I'm sorry to have to bother you with this. We are getting requests for interviews from all the TV stations and newspapers, and we would like to set up a press conference. It's absolutely your choice as to whether you want to talk to them. You can always have your daughters stand in for you, if that's what you'd prefer."

The girls glance at one another, exchanging nervous looks.

"What do we say?" Brandi asks.

"How many people will be there?" Angela wonders.

"What if we start crying?" Shelli says, with tears already in her eyes.

Damon assures them that he will do all he can to help them get through it. He tells them that he's already spoken to Jim's daughter, Laura, who has agreed to appear with them. Knowing that I can't do it, they walk through their fears, and don't even think of saying no.

Damon tells them he will schedule it for later that afternoon. As he turns to leave, he thanks them for doing this. Once he is out the door, the girls turn quiet for a few minutes. Each of them is lost in her own thoughts. Finally, Brandi speaks up.

"I think I can handle this," she says. "If you guys will sit with me, I can talk to the reporters."

As upset as she is, Brandi feels like this is a good time to do a few things she relishes: defend her mother, hold court before cameras, and talk nonstop. Having studied communications in college, she feels somewhat prepared. The story she is about to tell, though, is not one that even a seasoned reporter could do easily. Not when it involves your own mother. Not when you are experiencing it firsthand. No amount of study can prepare you for this.

Shelli is certain she'll burst into tears and isn't too thrilled about doing it with cameras rolling. Angela and Laura would rather keep their emotions private. They agree that Brandi will do the talking, although, truth be told, the girls don't know exactly what she's going to say. All they know is what they can see of our injuries. They have no idea how to begin answering the who, the how, and the why.

My boss, Brian, who has been here both days, offers to help the girls get through the press conference. Being in the cable business, he thinks he can give a hand. It brings me some relief that he's stepping up to give my girls support when I cannot. I try to talk Brian out of staying late, knowing he has a long trip home. He won't hear of it.

Later in the day, Damon arrives to escort everyone, except Jim and me, downstairs.

"Wish us luck, Mom," Brandi says to me as they leave my room.

I know they can hold their own, but I'm worried about them. They are tired, scared, and still in shock. I hate that they have to do this for me, but I'm proud of them.

While the press conference is under way, Detective Sparks comes back to my room for another interview. The local police are still keeping their ears and eyes open, but the case belongs to the Oklahoma State Bureau of Investigation.

"Cindi, I know your girls are talking to the press. That's fine. But we're going to keep this suspect information quiet for now. We don't want to tip him off."

"You already have a suspect?" I ask, surprised at his quick work.

"We have one in mind. Newcastle is not a big town. This kid is seventeen years old and has been in trouble before. We're going to question him. Maybe do a lie detector test."

Seventeen years old. His age shocks me. *He's a boy. He's a child. What would make a child do something like this?*

"Unfortunately, I can't tell you anything more yet. We're working on that. Our best evidence so far is the acid. We got enough of it from the scene of the crime to send it off to a lab to be tested," Sparks says. "We also found bits and pieces of the glass jar and a sizable chunk of a yellow lid. We're trying to determine where it might have come from."

A glass jar with a yellow lid. The description jogs my memory.

"Does the lid have ribbing on it?" I'd seen one just a week ago. I almost bought it, thinking I might make sun tea in it. I thought it would be perfect for summertime. To think, what if it were the same jar and someone else thought to use it to kill me?

"As a matter of fact, it does," Detective Sparks continues.

"You need to check out the nearest Dollar General store," I tell him.

This catches his attention. The nearest Dollar outlet is right there in Newcastle.

Will this tiny detail help unravel this horrible crime? I can't imagine that it will. I begin to realize how tedious the process is going to be. No fingerprints, no witnesses, no arrest. Just pieces of a jar, a bit of acid, and a shaky suspect. *Oh my God. This is not something I need to be worrying about. It is so far out of my control.* Talk of the investigation exhausts me like nothing else.

"Cindi, you look like you could use some rest. I'll check back in with you when I find out something more substantial," Sparks says as he stands to leave.

In my room, I lie there wondering how my girls are holding up. I am so proud of them for facing the press, for their willingness to stand in for me even though I know it's so hard. That's the risk of having such big hearts—the hurt is so deep.

The girls are all talking at once as they come back into my room.

"That sucked," says Brandi.

"You were awesome," Kendra reassures her. "Cindi, you would've been so proud." She puts an arm around Brandi. The girls are blowing their noses and wiping their eyes, trying to get themselves together.

"Tell me the whole story," I say.

They explain that on the elevator ride down to the first floor, they didn't say a word to one another. As they rounded the corner of a long hallway, they saw the windows of the hospital library. The press was already there.

"God, there were so many of them," says Shelli. Her voice is still trembling. "When we walked in the room, everyone stared at us. It was so surreal."

Dozens of people had squeezed into the small room. There were several lights and cameras and a mass of microphones on a long table. Damon helped them take their places and showed Kendra and Brian where to sit.

"We were so glad they were there. Two familiar faces made a big difference," Angela says.

They looked around, feeling overwhelmed. The four of them versus all of those strangers.

Damon introduced the girls, including Jim's daughter, Laura, who'd driven up from Texas. For quite a few seconds, the room was quiet.

"I just sat there biting my lip," Shelli says.

"My stomach hurt," says Angela.

"I started to get sweaty palms," Brandi says.

Finally, a reporter broke the silence simply by asking what had happened.

Despite the plan for Brandi to take charge, she froze.

"I know you're not going to believe this, Mom, because you know I'm never at a loss for words, but if I'd opened my mouth, I would've burst out crying."

The other girls say they held their breath, looking at Brandi, praying she had the strength to speak. Finally, Brandi did.

"All I could get out was that Angela had talked to the police officer, and he told her that Mom had been in an accident. It was too painful for me to say anything more. I just couldn't stop thinking about you up here, how you're suffering even more today than yesterday, how unfair it all is." Brandi starts to cry again.

In the pressroom, through her tears, Angela described the situation—how I heard an explosion and then grabbed my face from the pain.

"I just told them that I couldn't believe anyone could be so vicious to do this to such a good person."

The girls noticed watery eyes among the reporters, too, as they took their notes. They seemed almost afraid to ask more questions so as not to upset my girls even further.

So they turned to Laura and asked how her father was able to get help in his condition.

"I told them what Dad told me, that it sounded like a sonic boom when the jar hit the windshield, and that he really didn't know how he was able to make it to safety. I'm still amazed that he didn't hit someone or run off the road.

"None of us can stop asking *why*. Why would anyone do this? Why would anyone do this to you? It's so distressing. I told them that they won't find two better people in the world."

The girls said the hardest thing was keeping their anger hidden from the press. They held back the rawest emotions, not wanting to regret anything they might say.

"But, honestly," Shelli says, "what kind of world lets innocent, kind, and loving people suffer like this?"

Then a reporter asked the girls if they thought we were targeted.

"I said what you said, Mom, that basically you and Jim were just at the wrong place at the wrong time," Brandi says.

Another reporter wanted to know if the police had any leads.

The girls shook their heads to say no, unaware that Detective Sparks was up in my room describing the only suspect. An OSBI spokesperson answered the question for them.

"Once we can determine exactly what the substance was, we might have a little bit of a clue about who might have been responsible for this. All we know is that it was acidic and it burned through clothing and skin.

"And we still don't know if it was actually thrown off the bridge on purpose, if they were aiming, or if someone threw a bottle out of a car as they were driving over the bridge. If it was intentional, that's assault with a deadly weapon, which carries a twenty-year sentence."

"I know it was so hard," Kendra says sympathetically. "Maybe it'll bring good leads to the police. At the very least, it'll keep people talking. At best, the guy who did it might confess. You never know."

"Yeah, but it makes me physically sick to think he's still out there, running free, while Mom and Jim are in agony," Angela says with disgust.

I don't care who did it. I can't care. I don't have the strength.

All I can think about is getting by one minute at a time. Any time I moan or whimper, the girls ask what I need, and do whatever it takes. They move my arms for me. Change the bedpan. Cut up tiny pieces of food and feed me like an infant. Needing help with personal hygiene makes me know I have to get better. I don't want my girls burdened with an invalid mother for the rest of their lives.

All of a sudden, I see Robin's head peer around the door. Although she'd visited yesterday, I hadn't been able to process what it meant to have my sister by my side. Seeing her in person again turns my whimpering into weeping.

"Oh, Robin," I cry out.

Her eyes well up with tears at the sight of me. Then I notice she's brought Phil and their elder son, Jay, too.

The girls greet them, and I can tell they feel some relief that someone else is here to help.

Phil walks straight to my bedside and gives me a reassuring pat on the head.

"You doin' okay?"

I don't want to cry around Phil. I want to be strong. "I've been better."

"Ya think?"

He laughs, and so do I. My brother-in-law has always had that effect on me, even in this condition. I didn't know I was capable of laughter.

My arms are covered with bandages, and as I try to reach a tie on my robe, it becomes obvious that even this basic act is almost impossible for me because of my tender, swollen, and shredded skin.

Immediately, Brandi comes over to help.

"That's okay, I've got it. I can do it," I say while a searing pain shoots through my entire body. I look up at Phil and see something in his eyes. I'd call it respect. I've seen it before.

I know I have Phil's respect for raising my girls on my own. I never understood why he thought that made me courageous; I just did what the girls needed me to do. But whenever he says it, I feel good about myself. It means that I will be okay. He has accomplished so much; being one of his heroes is something special.

"Cindi, let Jay and me take the girls to lunch and give you some rest," Phil suggests. I never want my girls to leave my side, but he's right. They probably need a break—away from the hospital. Robin decides to stay behind and sits in a chair near my bed. I want to visit with her, but the morphine is making me so sleepy. Almost before they walk out the door, I briefly drift off.

When I wake up, I notice that Robin hasn't moved. She's still right there where I need her.

"Are you okay?" she asks.

I start to cry again for a different reason. Robin is one of my safest places, and I just let go.

"I'm so scared. I don't want to live the rest of my life incapacitated and dependent on the girls."

"Cindi, you're one of the strongest people I've ever known. There's no way you'll ever be a load on your kids." Robin's assurances mean a lot to me. I want so much to believe her. "You know we'll always be here to lift you. And there's someone pretty special who's thinking of you."

I can't imagine who it is. "Really?"

"Oprah called my cell phone while you were sleeping. She wanted to tell me how devastated she is for our family and especially for you. She asked how you are doing. I told her I wished you were awake so she could tell you herself. I tried to wake you because I knew you'd want to hear from her."

It's not surprising that Oprah called Robin, because of their close friendship. I'd met her once during her trial, and she couldn't have been kinder.

"I can't believe I slept through a call from Oprah, Robin."

"With all that morphine, you wouldn't have remembered it anyway." Robin smiles at me, and I doze off again.

As I slept, Phil took the girls to their favorite restaurant, Ted's Cafe Escondido, to grab a quick lunch, and they peppered him with questions.

"Is Mom going to be okay?" Shelli asked.

"You've got such a strong mom," he answered. "She's going to be more than okay."

"What should we do for her?" Angela knows what she can do as a nurse, but she wondered if Phil might have a prescription from the psychologist's point of view. He did. If there's one thing certain about Phil, he's never short on advice. And the girls trust him implicitly.

"Let her keep being a mother. I know your instinct is to want to take care of her, and that's fine. You can do that now. But she needs to be made to feel like she's still able to take care of you as soon as she's on the mend."

"So it's okay to keep asking her for money for gas?" Brandi couldn't help but crack a joke, but Phil took the point seriously.

"Actually, yes, if that's what you've always done. She needs to feel her life is still normal with respect to you girls. Help her, but don't make her feel like an invalid. Don't treat her like a victim."

When the girls tell me about their conversation with Phil, I realize that with his help, I don't have to worry so much about my daughters while I'm in the hospital. They'll get everything they need to make it while I'm out of commission—love from family, support from friends, and counsel from Phil. Knowing they are taken care of lifts an enormous burden off of me, and it's a great gift Phil gives me today. The other is this: because he is concentrating on finding the "bastard," I don't have to.

⁓

On Thursday, two days after the attack, Damon comes back into my room with another request from the media. He explains that every outlet in town wants to hear from Jim and me.

"We can't fit them all in your room, but I can arrange for representatives from TV, radio, and newspaper. That would mean about five reporters and photographers. I can keep it short if you think you're up to it."

Jim isn't crazy about the idea, but he leaves the final decision to me. I'm in too much shock to think too much about it. It seems easier to say yes than no.

Two hours later, they arrive and cram their equipment into our room. They ask a lot of the same questions my girls already answered, but Jim and I give them more detail about the moments leading up to and after the attack. I look straight in the eyes of the newspaper reporter. He is young and handsome, and his looks are such a contrast to mine. Seeing the sympathy on his face, I know I look terrible.

The interview is short, just as Damon promised. A photographer snaps a few more pictures. All of them are very polite as they pack up their equipment and wish me well.

After they leave, the whole experience feels surreal. Thinking about why they were here is exhausting.

I'd rather spend time thinking about the generosity of all the people who love me, and barely a second on my attacker. I do want the OSBI detective to find him, and let the justice system punish him. After my days in the hospital, I created my own brand of punishment: force him to spend time in this burn center where he would hear the screams and cries of burn patients until he felt agony over what he did.

It distresses my kids and friends that I'm not intent on a worse punishment. Sometimes I wonder if there is something wrong with me. But I don't want my life to be consumed with hate. I think how sad his life must be to have done this to a totally innocent stranger.

He couldn't be a happy person. If he did it for a thrill, he's pathetic. He has no conscience. No soul. I keep thinking, Maybe if someone had done something small, simple, and sweet for him, a tiny act of kindness, even a smile and a *How are you?* that night, it might have changed his mind.

Maybe I'd feel differently if I saw him. I know I'd feel terrible if he smirked or laughed or just didn't care. That would break my heart—to see no sign of remorse. I also know that the choice to forgive is one each of us has to make. Admittedly for me, this would be far more difficult if one of my daughters or grandchildren had been attacked. Fortunately, I have not been faced with such a situation. But as for me and my life, forgiveness has been a powerful part of my recovery. Refusing to be angry at a person I do not—and may never—know gives me peace.

In my nightmares, I see the saddest set of eyes. They are brown and lifeless; they have no feeling, even though they are the eyes of a child with a lifetime ahead of him. He chose to commit a single act of violence, over and done with in an instant, with an end result that will last for the rest of my life. Every day from now on I will wear the scars of the hatred that was inside of this person. I will live with the visible marks of his pain. I will endure endless stares and unkind remarks. I will answer the questions of small children and adults alike. And I do worry about him hurting someone else; that's the only reason I really want him caught. But I haven't wasted my energy on that, and I'm not going to start now. He'll have to live with painful guilt associated with his bad choice that day, and his maddening choice to stay silent since. Guilt, if he feels it, may have to be punishment enough.

People may forever be focused on discovering the identity of my attacker. My daughters, my family, Phil, the detectives, and even

Officer Norman are all fighting that fight for me, so that I don't have to. They allow me to devote my life to healing: healing my scars, healing the hurt in the world, healing the open wounds of people like the one who dropped that jar of acid. I choose to let him go. Instead, I will use my strength and positive thoughts toward ends like getting out of the hospital and getting on with my life.

If I don't, he will burn me over and over and over again.

I won't let him win.

As I come to the crossroads of retribution on one extreme, and forgiveness on the other, I choose the other—completely. Forgiveness is not a gift to him; it's a gift to myself. Even though his choice put my body through living hell, *my* choice to forgive gives my life back to me.

Vengeance is not mine; I will let God handle that.

My inspiration: one of the photographs of Kennedy and me
that I had taped to the foot of my hospital bed

5

Humor Doesn't Hurt

EVEN THOUGH I'VE HANDED vengeance over to God, I'm the only one who can handle the assault on my body over the next sixteen days. It will take everything I have to survive these battles, even a sense of humor—an unlikely but powerful ally I don't expect to find during these difficult days spent in the burn unit.

"Good morning, Cindi. I'm sorry, but it's time," the nurse says as she pushes a wheelchair into my room. Her voice is cheerful, but I've come to dread it.

Each morning I wake up, get a dose of morphine, and then, accompanied by one of my girls, head to the tank room for daily, sometimes twice a day, debridement. It's where nurses remove dead skin to help the healing process along. As you might imagine, there is no gentle way to take skin off. I am one giant blister that has popped, exposing nerve endings. *Not* touching them hurts. Rubbing them raw is unbearable.

Slowly, the nurses lower me onto a metal chair under the tank room shower. I sit stark naked and wait for that first sting of water to hit my tender skin. This agonizing ritual of cleaning, scraping, medicating, and then wrapping my wounds in fresh bandages doesn't get any easier. I can't help the tears. The nurses know I hate going in there, and they claim I'll feel better when I come out. *Better compared to what? Having my nerve-exposed skin rubbed with a Brillo pad versus being left alone?* They are very aware of my pain and do everything in their power to minimize it. But some things in life are just not easy no matter what is done. This is one of those times.

"Ohhhhh," I moan.

"I'm sorry, Cindi," apologizes Jessica, a young nursing assistant who bravely asked to be my tank room nurse. "We've got to keep your wounds clean so you don't get an infection."

The staff at the burn unit has a special sense of sympathy for me because of the way in which I was injured. Most of their patients have some forewarning that they could be hurt. Whether it's an industrial worker handling caustic chemicals, or a drug addict whose methamphetamine brew explodes, the patients knew there was a possibility they could be burned. I can't imagine what those nurses and doctors go through—pouring out all their expertise to save the life of a meth addict who nearly blew up the neighborhood. Nonetheless, they don't hold back treatment just because someone is a felon.

My case is different because the attack was so random, so unfair, so evil. There was nothing I did to provoke it; nothing I could have done to avoid it. The medical staff all know this. But having their sympathy doesn't translate into a less excruciating experience.

"Ohhhhh." It's all I can say.

I know about the expression "What doesn't kill you will make you stronger." I cannot imagine how being debrided will strengthen me, but it has to be done. Just when I nearly faint from the pain, it's over. I'm so exhausted that I can't wait to get back to bed. Jessica wheels me back to my room and helps me settle in.

"Cindi, let me know if you need anything, okay?"

I nod and feel my eyelids close. *Thank God it's over.*

But it's far from over.

Each day it becomes clearer to the doctor that I'm going to need skin grafts. This afternoon, he breaks the news to me and my daughters.

"Cindi, we all hoped your burns wouldn't warrant them, but skin grafts are unavoidable. Your burns are simply too large and too deep to heal on their own."

"When will we do the surgery?" I ask.

"We can't do it right away. You'll have to undergo debridement first."

"But I just did debridement this morning."

"No, not that kind. This is a surgical debridement done under general anesthesia. It's quite a bit more intense. I have room in my schedule to do it this evening."

More intense. I can't even imagine what that means. This will be my first surgical procedure. A major debridement.

"Why does she have to go through this surgery?" Brandi asks.

"These burns are not leaving a smooth, level surface. We have to create one by cutting away the dead tissue down to living skin."

I can hardly believe the doctor is talking about my body. All I can liken it to in my mind is cultivating a garden. You don't want to spread seeds over dead weeds. You have to cut away the dead growth to make way for the new. You have to till the soil so the seeds can take root. Getting ready for skin grafts sounds like the same thing. The grafts won't take well if the skin isn't tilled. But the earth isn't filled with millions of nerve endings.

The girls and I are in shock. We had hoped I'd be going home the next day with Jim. Now our hope has turned to fear.

Shelli asks Angela, "Is she strong enough to make it through two surgeries?"

Nothing I say can comfort my daughters. It's clear they are deeply worried. Although I'm trying not to let them know, I'm scared to death, too.

Before we know it, the time for the surgical debridement arrives. The girls follow me downstairs to an area just outside the operating room.

"That's as far as you can go," the nurse tells them. "It's going to be several hours. You can wait right over there." She points to a small empty waiting area as she takes me down the hall.

In a strange way, I consider myself the lucky one. I'm going to sleep, unaware of anything after this point. My girls, on the other hand, are miles from home and their families. They are left to wait and worry with only each other for comfort.

I don't remember much about going into or coming out of the surgery. Under the spell of heavy painkillers, I have no idea that two hours pass before the whole thing ends. I have a vague notion that an OR nurse is leaving me in recovery to go talk with the girls.

Two hours later, as I am wheeled out, I hope that the nurse gave the girls some warning. I must look far worse than they imagined, because the sight of me takes their breath away. I hear their gasps. Shelli literally falls to her knees. Brandi has to find a place to sit. Even Angela, who has seen some terribly gory surgeries in her years of nursing, has never looked so horrified. It's difficult to see their mother under all the blood.

"My God, it looks like they skinned her," Angela says out loud.

"That is the grossest thing I've ever seen," Shelli whispers on the floor.

My skin looks like raw hamburger meat—pink, mushy, and bleeding from various folds. My face is pooled in blood. The rest of my body is wrapped like a mummy. The layers of bandages are already saturated with blood.

The girls tell me that this is the most traumatizing sight they've ever seen—one they will never forget. Normally I'd be the one to console them after such horror, but more than ever, I'm powerless. We all go to sleep this night knowing that our nightmares can't be any more frightening than what we'd just experienced.

But the worst is yet to come.

THE NEXT MORNING, I wake up, my body still sore from surgery and my mind still fuzzy from morphine. As my vision comes into focus, I find myself gazing at the photographs of my granddaughter Kennedy taped to the foot of my bed.

Kennedy and I haven't laid eyes on each other in nearly a week, which leaves a gaping hole in my heart, and lot of confusion in her head. It's not normal for us to go more than a day or two without a visit, and it hurts to be separated from her. She is my jewel, my pride, and my Tootie Girl, the first to usher me into the wonderful world of grandparenthood.

I look over at Angela, who has just come back into the room from breakfast.

"When do you think you could bring Kennedy up here?" I ask.

"I'm not sure, Mom. I know she really misses you, too."

Angela has to hear the question from both of us, two people she'd like nothing more than to please. She's caught in the middle of a difficult dilemma—protecting her daughter from potential harm while giving her mother whatever she needs to heal.

"You know, Mom, I'm not even sure if children are allowed up here. But I'll look into it," Angela fibs. I have the feeling that she tells Kennedy the same thing.

Essentially, Angela puts us both off as gently as she can. She's torn. Angela and Steven know how much Kennedy and I love each other, but they must worry about whether Kennedy should be exposed to such brutal reality—the burns, the pain, the crime. *Will she understand any of it? Should she see any of it? Will the sight of her grandma lying in a hospital bed, burned and bandaged, unable to give a hug, much less a smile, devastate her?*

For now, I have to accept Angela's answer. There's something more pressing on my mind. I have to get up. Cautiously, trying not to intensify the pain I feel everywhere, I begin my first odyssey out of bed toward an ambitious destination—the bathroom. I barely move an inch, and already it feels like every bone in my body is hitting against the other. Ever so slowly, I slide my legs over the side of the bed and take a deep breath. Immediately the girls jump up and position themselves on either side of me. One of them grabs a walker and pulls it to the edge of the bed. I am sure they do not know what to think. Mother Nature is calling, and I am determined to make the six or seven steps to the bathroom. They are just as determined that no more harm comes to me.

Each movement becomes a small victory in this, my first foray out of bed. But by the second step, my body starts to shake violently. I look so pathetic, covered in blood and practically seizing. My jaw starts to hurt from shaking so much.

"Mom, why are you shaking like that? Are you sure you're okay?" Shelli's voice is panicked.

By the time I make it to the toilet, I'm shaking uncontrollably, but the answer to her question comes in the form of a question of my own.

"Could it be just maybe I've been in a bad accident?" I say sarcastically and shake out a giggle. One by one, the worried looks on my girls' faces soften. I surprise myself by answering the way I might have done before this ordeal. Acid burned through my skin, but not my wish to calm my girls and prove I still have a sense of humor. In our house, Brandi is the queen of sarcasm, so what better moment to dethrone her? It turns out I'm chilled because I've lost the protective covering of my skin, along with a lot of blood. Laughing with my girls warms my heart, and I feel my shaking subside.

I was so focused on making it to the toilet that I hadn't noticed the bathroom mirror. But as I shuffle back to the bed, I catch my reflection.

I stop to take a look.

Oh my gosh, can that really be me? It's no wonder people look at me with such pity. I am indeed a very pitiful sight. Suddenly, a lot of things become clearer to me. The brutality of it all rears its ugly head, so to speak, right here on my face. Because I hadn't been able to see at first, much less get up to look in a mirror, I haven't known how horrific it is.

My new face is a terrible patchwork of yellow, red, purple, and brown, colors I never knew skin could turn. It has swelled even more today. I can't open my mouth. I look like the Elephant Man. I can't tell what's making my face so shiny: the ointment coating my skin or pus.

Still staring at my face, I take a deep breath and slowly exhale. The room is deafeningly silent. I may be the only one breathing.

I am so stunned by my own face that it takes a few seconds to register what I'm looking at. I turn my head from side to side so as not to miss anything. So many thoughts are racing through my mind, but the only sound I can make is a laugh.

It's tough to tell that I'm actually laughing. My face doesn't move. At first the girls think I'm crying, but then they recognize my chuckle. In an instant the room becomes a flurry of smiles, movement, and conversation, all signs of relief. The girls can breathe again. They want to protect me, but they can't keep my eyes closed for me. It's one worry they can release.

"You need to get back to bed now, Mom," Angela says as she ushers me away from the mirror.

"We promise you, it's not as bad as it looks," Brandi says.

Now that's funny, because it is every bit as bad as it looks, and it may be getting worse. I'm touched by the way the girls seem desperate to say anything to make me feel better. I almost wish Brandi and Kendra would do one of their stand-up comedy routines, but the girls aren't comfortable enough yet to make me laugh that hard. I have a lot more hell to face, and they are courageously walking the walk with me.

Watching my girls arrange the sheets around me, all I want to tell them is, *This does not hurt me as much as it hurts you. I'm so sorry that some stranger has caused you to feel so much terror. I promise you someday I will make this up to you. I will do whatever is in my power to make the world a kinder place for you to live in. I vow to show kindness in response to the cruelty shown to me.*

Before I'd seen my reflection, I hadn't *truly* realized what Angela must be thinking each time I ask to see Kennedy. But now that I've seen my face, I see the dilemma I've caused, and I won't put Angela in that position again.

It turns out I won't have to.

IT'S SUNDAY AFTERNOON, just five days after the acid attack.

"Mom, would you mind if we left for a little while?" Angela asks sweetly, even though she doesn't need to. Steven and Angela have spent so many hours cooped up in the hospital with me that they decide Kennedy needs some time with them doing something entertaining. I encourage the idea, although all I want is to have Kennedy right next to me.

"Of course not," I answer. "You need a little getaway, too. But I want to hear about how much fun you have."

So they pack up the car and drive to Tulsa for a birthday party. Ninety miles by car is a long way to travel for a party, but they figure Kennedy could use a joy-filled distraction.

I later learn that on their way back from Tulsa, Angela asked Steven to drop her off at the hospital so she could resume visiting with me. Once they arrived in the parking lot, Angela turned to tell Kennedy good-bye.

"Where are you going?" Kennedy asked.

"Mommy's going to see Grandma. This is where she's been for the past few days." And then, without even thinking, Angela said, "You know she wants to see you."

Kennedy looked at her mother in silence. Angela continued, measuring Kennedy's reaction to every word.

"The thing is, Grandma doesn't look like Grandma. Her face is red and swollen. But even though she doesn't look the same, she's still the same Grandma on the inside. She doesn't feel very well, but seeing you would make her very happy. It's up to you. Do you want to see her now?"

Kennedy looked up at the burn unit towering over the parking lot, and then turned back to Angela.

"Tell me that story again, Mom."

Angela wasn't sure what to do, except to simply repeat what she had said.

"Grandma doesn't look like Grandma. Her face is red and swollen. But even though she doesn't look the same, she's still the same Grandma on the inside. What do you think?"

It was as if Kennedy sensed her mother's words were so serious and so important that she had a duty to fully absorb each one. She wanted to study before the test. She needed to understand the truth intellectually before facing it emotionally.

Angela tells me she thought Kennedy might have said she wasn't ready. Something like, *Not today, Mom.* But Kennedy sat silently in the car for a little while longer.

"Tell me one more time."

Word for word, Angela patiently repeated the story. The third time around was indeed the charm.

"So, do you want to see her now?" Angela asked again.

"Of course I do. Can we go now?"

And that's Kennedy. She had the facts. Once it was set in her mind, it made perfect sense.

Angela comes into my room with a refreshed smile on her face. It looks to me as though the party did her some good.

"I have a surprise for you," she says.

"Oh, you do?" Most times, a surprise would really thrill me, but I'm not expecting anything more than a bunch of flowers. I don't want to seem ungrateful, but whatever she is bringing me can't take my mind off my misery. I feel like I've had enough surprises for the time being.

But her smiling face should've told me it's more than gift-shop flowers.

"Kennedy is downstairs."

This, I do not expect. "Ohhhh good. I get to seeee herrrr!"

My voice is high and quavering, which happens when I talk and cry at the same time. Instantly, my whole attitude changes. My heart beats a little faster. I feel like the weight of the world is lifting off my shoulders.

I turn to Shelli and Brandi. "Do I look okay?"

"Mom, you look great," Shelli says.

"Steven is going to bring her up," Angela continues. "First, I want you to know what I told her." Angela explains what had gone on in the parking lot, and how confident Kennedy seemed in her decision to come see me.

My eyes are fixed on the door, almost unbelieving that my grandbaby will come walking through it.

The closest I've been to Kennedy over the past horrifying days are those photographs taped to the foot of my hospital bed. Every time I look at our faces, turned toward each other in one shot, and then toward the camera in another, I can hear the laughter we shared that day. I think about all our good times—but especially the night before the attack, walking back from Kennedy's T-ball game, without a care in the world, without a clue that life as we knew it would be utterly transformed come morning.

Before too long, I see Steven. In one hand he holds a small box, and in the other, Kennedy's little hand.

Just as she appears next to her father, Kennedy stops in the doorway, and my heart stops in my chest. Suddenly, adrenaline does what morphine cannot; I feel nothing, except deep love for my granddaughter. She looks straight at me, takes a deep breath, pries the box from her father, and walks over to my bed.

"Hi, Grandma. I brought you a piece of cake."

It turns out, during the party, when slices of birthday cake were passed around, Kennedy politely asked if she could have hers wrapped up to eat later. The hostess complied, putting the chocolate frosted cake in a colorful box, similar to one you'd find at a Chinese take-out place.

"And I have to tell you something."

There's no telling what will come next, but I'm pretty certain it'll be precious.

"One time I had a red booboo, and my daddy put medicine on it and made it all better."

All of us burst out laughing. The burns on my face remind her of the heat rash she once had on her rear end.

"Good advice, Tootie Girl." I look at Kennedy's dad and say, "Do you have any more of that stuff?"

He smiles back at me. "Not that much!"

We laugh even harder.

Then Kennedy takes over. She talks about the party, the presents, and how much she's missed me. Not for one second does she act like anything has changed between us. She pays no attention to the yards of bandages, my scabby face, or the fact that Grandma can't reach down and pull her up onto my lap as I always do. Any fear that I had about whether Kennedy would find my swelling, scabs, and sores frightening disappears. Children are amazing. So often, when they speak the truth out loud, it's an embarrassing episode. But this time, Kennedy's words mean she sees past the exterior to the truth, and it is a great relief to me.

Kennedy reaches over and picks up an odd-looking contraption on the nightstand.

"What's this, Grandma?"

"It's my self-scratcher. Your mom made that for me to scratch my back."

"This is weird."

"I know, Tootie Girl, but I'd go crazy without it."

The skin around the burns itches terribly, seemingly all the time. The frustrating thing is I can't reach the places that itch because my arms are wrapped up so tightly. I hate to ask the girls to scratch them for me all day long. Watching me struggle drives them mad. So Angela created the self-scratcher.

As Kennedy inspects the odd-looking thing made of straws taped together and attached to a small sponge, I realize how funny—and weird—it truly is. What I don't tell Kennedy, to spare her from disgust, is that the sponge is not supposed to be for the parts of my body that itch; it's for my tongue.

I ingested enough acid while screaming for help that the inside of my mouth is also burned. My tongue feels six inches thick, like I might choke on it if it swells any more, and is starting to peel off in layers. I get the job of scraping it out with a special sponge to keep from gagging on my own dead skin. This becomes a ritual that even the most experienced nurse cannot handle, so I'm on my own. I decide that my four-year-old granddaughter doesn't need to know all the gory details.

While Kennedy and I are visiting, Angela pulls Steven aside.

"Steven, I need you to do something for Mom."

"Sure, anything."

She makes her request quietly so no one else can hear her.

Steven shouts back, "You've got to be kidding me!"

This gets all of our attention. Angela tries to tell us what she's asked of him, but she's cracking up. Finally she says, "I asked him to go buy some new underwear for you."

One of the first things Robin did for me was to bring soft gowns and comfortable, even fashionable, robes for me to wear around the hospital. *If I am going to be here half naked for some time, I might as well do it in style.* But she didn't know that I needed underwear. Acid had ruined my underwear in the attack, and I can't stand going without it. Even though the medication should soothe me and the pain distract me, I really feel uncomfortable without a good pair of panties.

"You're going to owe me big-time for this," Steven says as he looks over at me.

"Size seven, whatever you can find," I tell him.

Of course the poor guy can't refuse, and I'm not thinking about how awkward he might feel buying underwear for his mother-in-law. Steven dutifully drives to the mall, goes into two different department stores, and proceeds to buy every pair of underwear imaginable.

He comes back with two big bags. Inside are pink, blue, purple, striped, floral, lace, cotton, satin . . . honestly, it's as if he brings the whole lingerie department to my bed. I notice that the only style he avoids is the thong. He's mortified but hides his embarrassment well. And for him it only gets worse. As the girls go through Steven's selections, Angela yells out, "LEOPARD PRINT, STEVEN?!"

We are all laughing hard, and the release feels so good. Finding humor in the smallest things lifts heavy burdens off our chests and helps us relax. Laughing together makes us forget, just for a moment, what our family is facing.

I'm coming to a second crossroads, entering the most painful chapter of my recovery. What choices would I make to help me get through it? Instinctively, I lean on laughter. Don't misunderstand—I scream and cry and moan far more than I laugh. But humor was a part of my happiest days, so why can't it be part of my most difficult moments? I know I have to do something to keep my spirits up, or I'll take everyone right down with me.

Just as I feel God led me to the place I needed to be, in one of the best burn units in the country, with a knowledgeable staff and state-of-the-art equipment to help heal my burns, he also gave the healing power of humor. Even if it's only for a few seconds, when I'm laughing I don't have time to think about the pain, the fear, or the situation I'm in. In trying times, finding humor can be difficult or even seem inappropriate. But I've found laughter is my best medicine.

This has been another incredibly long day.

My emotions have gone from one extreme to the other. I've cried and I've laughed. I've been in terrible pain, and I've forgotten all about it. I was thrilled to see Kennedy but deflated when she had to go. I'm happy for Jim because he's going home tonight, but I'm also sad because he's leaving.

Jim comes into my room to say good-bye. He will have to come back frequently for bandage changes, but still, it won't be the same not having him close by. For the first time since the attack, our paths diverge. We are being separated, left to face our fates without each other.

Before Jim leaves, we have just a few minutes to ourselves. He sits in the chair close to my bed and takes my hand in his. Neither of us speaks for a few seconds. Knowing the kind of person Jim is, and the loyal boyfriend he has been, I imagine he's feeling guilt at having to leave me here.

"Jim, remember, I think you're a hero. You saved my life. It's amazing to me—how you kept driving, how you found the showers, how you completely ignored my sweet request to pull over."

"Cindi, I'm no hero. You were screaming. Screaming bloody murder. I have no idea how I kept my head through your screams," Jim says, shaking his head. "I have no idea what kept me going. Instinct, I guess."

"No, I think it was something else," I smile again as the thought forms in my mind. "I think my mother was there, sitting on my shoulder, telling me I was going to be okay. And my dad was on your shoulder saying, 'Drive, you son of a bitch, drive!' "

We both smile, relieved that we are alive, that we can still laugh together. And though my parents are both gone, a glimpse of my mother's comfort and a bit of my father's humor make me feel that in spirit, they are with us.

Jim leans over and kisses the top of my head.

"I'll be thinking about you tomorrow during your skin graft surgery."

As Jim starts to walk away, he turns to face me. "And Cindi, as far as I'm concerned, you're the hero."

I tell him, "You always have to have the last word."

We both laugh, and I can hear him chuckle all the way down the hall.

6

Going Home

IN THE HOURS after skin graft surgery, there is no room for laughter.

I wake up screaming. My entire body is on fire. Something must be killing me. This is the worst pain I've ever felt in my life. Worse than the acid, worse than the debridement. I beg for my misery to end. I beg for my life.

"Angela, please don't let me die!"

When the doctor said I would need skin grafts, I knew they would take part of my skin from one place and put it on another. But I had no idea my upper legs—the only areas healthy enough from which to harvest skin for the grafts—would hurt worse than my original burns.

"Somebody get her some help!" Angela demands. Up to this point, my girls have felt like they could help me in some way. But it's clear this is far beyond their control.

As Brandi runs to the nurses' station, Angela is pumping the call button. Shelli paces. The girls cannot handle seeing me in so much anguish.

Home at last!

I cannot even begin to describe the torture that skin grafts bring. There are more than four hundred staples holding them in place.

Four hundred staples.

I have grafts on my upper eyelids, cheeks, chin, neck, forearms, chest, and from the knees down on both legs. They removed enough healthy thigh skin and ran it through a mesher to expand it wide and long enough to cover 180 square inches of my burned body. Even the head surgeon tells me he's surprised at how much area must be grafted. But they have no option because of the way the acid splattered on me. It's too tedious to surgically repair the damage made by every little drop, so they have to cover whole areas of my body.

I can live with that. I can live with the staples, the meshed and grafted skin. But I'm not sure I can survive these donor sites on the front of my thighs. From my hips to my knees, that good, healthy skin has been sliced away, leaving exposed nerve endings. Doctors took five strips of skin from my thighs, each four inches wide.

"My legs are burning!" I scream.

Angela places her hands an inch above my legs.

"Shelli, come here and feel this. You can literally feel the heat radiating from her legs."

Shelli is shocked. "My God, it feels like a hot stove."

I now question whether I have the strength to continue. I can't stand the pain any longer. I've hurt as much as I can, and it's far worse than anything I've gone through. For the first time I wonder if death might just be easier. But I quickly force that thought from my mind. It's not even an option, just the pain talking.

The girls are in tears. It broke their hearts to hear about my screams when the acid started burning us. It's tearing their hearts out to witness it firsthand.

Angela takes my hand. "Mom, I promise you won't die. I'll get you help."

Brandi comes into the room with a nurse in tow.

Angela turns to the nurse. "You've got to give her some more pain medication."

"She's maxed out. If we give her any more, she'll stop breathing," the nurse argues.

"Then why wasn't she knocked out a little longer? You can't just let her lie here like this." Angela's tone turns more forceful. "If you won't do it, I want to talk to the doctor."

"It's too late to call. I don't want to bother him with this."

With that, all three girls launch into protective mode, share a few choice words of their own, determined that someone is going to listen and respond.

I hear arguing and desperation in my girls' voices. From my bed I call out again, "Am I dying?"

Angela can't take it anymore. "Either call that doctor now, or all hell is going to break loose."

The nurse makes the call, and Angela convinces the doctor that I need more morphine.

It's the worst night I've ever had, and the beginning of the longest ten days of my life.

∽

THE NEXT MORNING I wake up in disbelief.

This cannot be happening. I thought I'd already gone through the hardest part. No one prepared me for this, although I'm not sure anyone could have.

I can barely breathe. Any movement brings on intense suffering. The fire in my legs is still raging.

"Off to the tank room, Cindi," a nurse comes in pushing a wheelchair. It must be Jessica's day off. Brandi looks at me with real skepticism. It's her turn to stay with me during debridement, and she would not have picked this particular nurse, who always seems to have a scowl on her face, for such a difficult morning.

"Are you kidding me?" Brandi jumps in. "She just had skin grafts. She's in so much pain. Can't she have some time to recover?" Brandi can't imagine I can take any more suffering.

"I'm not going to touch the grafted areas for a few days. But her donor sites need to be cleaned."

"No! I can't take that."

"Mom, I'll be right there with you," Brandi assures me.

The nurse shakes her head. "Not this time you won't."

"What? My sisters and I have been with her in that tank room every single day."

"I'm going to have to insist you stay here this time," the nurse says, almost glaring at Brandi. It's becoming a battle of wills.

I can tell Brandi is beyond upset. Her eyes narrow as she looks right at the nurse. "Well, I'm all she's got right now."

"Beebop, it'll be okay. I can handle it," I say. I'm not sure I can, but I want to avoid a brawl if possible.

As Brandi starts to storm out of the room she says under her breath, "Well, she doesn't have to be so frickin' rude about it." She stomps off down the hallway.

The nurse looks right at me. "Well, if she isn't a banty rooster."

I don't go into the tank room smiling, but I can't help but think,

There goes my little banty rooster. Stubborn, maybe. Tenacious, defi-nitely. Thank God she's on my side.

The only way I can describe postsurgery tank room debridement is to say it is torturous. If I thought it was bad before, I now realize it was a cakewalk compared to this. The nurses try to be more gentle than usual, but I understand why they don't want my family in here with me. At least it's a short visit, since they're cleaning only my legs.

I'm nearly unconscious when I am brought back to my bed. Brandi is there waiting for me.

I look down at my thighs, covered in wet gauze. On top of ex-cruciating pain, it feels like I'm lying in pee. I press the button for another dose of morphine. It allows me a few hours of restful sleep.

As I wake up, Brandi looks like she's calmed down. She tells me about the visitors who dropped by and the phone calls she took. The nurse who'd taken me to the tank room comes in to check on me. She leaves without even looking at Brandi.

I turn to Brandi. "She thinks you're a little banty rooster."

At first Brandi looks offended. But then she thinks about it, more perplexed than anything else. "Well, is that a good thing or a bad thing?"

"The nurse clearly means it as an insult, but I can't help thinking about my daddy who had a banty rooster on his farm. That bird nearly drove him crazy. The truth is that you, he, and the rooster were all equally feisty, determined, and stubborn. Basically a pain in the butt."

Brandi smiles, deciding that the nurse's characterization of her is really a badge of honor.

"I could think of a better name to call her than that," she smirks.

A few minutes later my doctor knocks on the door as he comes by on his rounds.

"How are we today?"

"This stuff is killing me," I tell him, pointing to the heavy bandages on my legs.

"That's not what hurts," he explains. "That's silver ion, a good barrier to bacteria. It's expensive too. Each strip costs a few hundred dollars. You'll come to appreciate it because it means the nurses don't have to mess with the donor sites as much. We can leave the gauze on for four or five days."

"Does it have to be so wet?" Brandi asks.

"That's what keeps the silver ion activated."

"When will I stop hurting?" I ask before he leaves.

"You have to remember that skin grafts are just like partial thickness, or second-degree, burns. They expose those nerve endings. Your third-degree burns don't hurt as much because they destroyed the nerve endings. Some patients say the grafts get better in seven days," he answers. "It could take ten."

As far as I'm concerned, the cure is worse than the disease.

~

NOT TWENTY-FOUR HOURS after skin graft surgery, I'm visited by a physical therapist.

Up to this point, I've been cooperative, doing whatever the nurses ask of me. My daughters have given me a lot of care and taken some of the workload off the nurses, but not necessarily the strain. Brandi and Angela have been going head-to-head with them for a few days now. The nurses make it clear that they view the girls' involvement as interference. Personalities are clashing.

And the tree doesn't stand far from the fallen apple.

As the initial shock of the skin grafts wears off, I become more aware of my pain, and less willing to comply with anything that intensifies it. Just lying in my bed is torment. I don't want to move, and yet it seems to me that after a few short sessions together the physical therapist expects aerobics.

Physical therapy is intensely painful. After daily debriding, a surgical debridement, skin grafts, and more debriding, exercising my besieged body feels like the last thing I need. And it's heart-wrenching for my daughters to watch.

The therapist pushes me through the pain, showing me just how she wants the exercises done. But as soon as she leaves the room, I give up. It hurts too much.

When the therapist comes back into the room to check on my progress, she's clearly annoyed.

"You need to get to work," she says.

"I can't," I say flatly. I'm irritated with her apparent lack of concern for my struggle and decide that I am going on strike for the moment.

I know the exercises are for my own good. They're supposed to help my skin become supple and hasten my healing. I don't care. It just plain hurts too much. My daughters continue to help me find every excuse not to bring on more pain. I know our united front drives the therapist crazy.

This particular therapist doesn't care too much for me. And I don't care for her. I think she's pushy, and she thinks I'm stubborn.

"I can't do this anymore," I tell her in the middle of stretching and bending my arm where the skin graft had recently been sutured. "It hurts too much. I'm miserable."

"You have to. You're going to be more miserable if you don't do your exercises."

As she leaves my room, someone stops her in the hallway. I can hear their conversation.

"That Cindi needs to take a lesson in attitude adjustment from her brother-in-law," she says with contempt.

What? How dare she! At first I'm shocked at being judged by a health professional who is supposed to have compassion for her suffering patients. Her comment bothers me all day. *What a witch.*

But the more I think about it, I realize she is absolutely right. I do need a lesson. And it puts Phil's voice right in my head. *This may be painful, but this is for* you. *It's not for the therapist or the nurses or the doctors. It doesn't entertain them to see you in pain. They know it's good for you in the long run. And you have to take care of yourself. This is a way you can help your own recovery. It's a way you can control whether you get better or you don't.*

Suck it up, you big whiny baby.

Phil only calls me "whiny baby" when he knows I'm really suffering. Time and time again I've heard Phil talk about doing what is painful now to ease pain in the future. Emotional choices like breaking off an abusive relationship or showing tough love to a spoiled child are examples I understand. But this is physical pain, and somehow it interferes with my ability to remember that lesson and put it into practice. The burn unit therapist shames me into remembering as I laugh at my own stubbornness. It's what Phil would call a defining moment in my recovery.

The next day the therapist walks in, and to her surprise, I'm doing those exercises on my own. I don't make eye contact with her. I simply concentrate on moving my left arm up and down, vigorously and precisely, working through the agony on my own. I think she almost faints.

"Whoa!" is all she can say.

She has no idea that I overheard her complaint, nor what power her words had on my attitude.

But we still can't stand each other.

⁓

I CAN SENSE THAT there is something else annoying the nurses: all the attention I am being given. I have more cards, flowers, and visitors than anyone on the floor. They fill up every table, counter, and inch of wall. There's hardly room for a nurse to take my temperature or check my monitors without having to ask someone to get out of the way.

All of it is critical for me.

But as the days pass, my room becomes less crowded with people. Jim is no longer in the room across the hall. Angela, who had to give up breastfeeding to be by my side, goes home to Carson and Kennedy. Shelli had to get back to her job as a hairdresser. Only Brandi, who had just resigned from her job as a customer service representative for a rental car company, and Kendra are still around. Being a prekindergarten teacher out of school for the summer, Kendra can stay with me nearly every day.

On the seventh morning after the skin graft surgery, the day the doctor said the pain from my donor sites might subside, I wake up eager to assess my pain. My surgeries in the burn unit are over. I haven't been able to walk since the skin grafts, but I can feel that I'm starting to heal. The pain in my legs is not gone, but it's not as severe as it has been.

"Kendra, I think we might be able to do something today."

Kendra smiles. "Great. What's your pleasure?"

"I'm bored with this room. I thought we'd decorate."

I say we, but Kendra does the work and lets me do the directing. There are piles of cards everywhere. A Sunday school class made cards and packed them into a huge manila envelope. A group of schoolchildren mailed drawings. Colleagues, friends, and strangers sent get-well wishes. Kendra picks up the biggest stack of paper I've ever seen and starts wallpapering every open space on the hospital room walls.

Kendra has done far more difficult chores: changed my bedpan, held a bucket for me to throw up in, and fed me by hand. Spending six hours putting those cards up, which is exactly how long it's going to take, is a picnic by comparison.

"Where do you want this?" Kendra asks with every single note, card, and drawing.

She stands back, waits for my approval, rips a little piece of white hospital tape, and attaches it open and flat on the wall so that I can see what's handwritten inside.

Sometimes I ask her to read them first. A lot of them make me cry. I have her put the ones that make me laugh closest to my bed.

"Karen and Mike write, 'I know you really wanted that facial laser peel but really, there was an easier way!' " Kendra reads.

"What about that big card?"

"It looks like one from your office. Mary says, 'What a way to get a free facial chemical peel. I'm so jealous I had to pay for mine! (And you actually got a full body peel!) You're a remarkable woman!' Then someone scribbled, 'She's one sick puppy' by Mary's signature. Look."

"Why don't you move that a little there, so I can read it? I like that one."

Morphine is probably keeping me from giving directions in an organized or timely manner. This process has to be maddening. But Kendra doesn't flinch.

"Okay, Cindi. What about this one?"

And so it goes, one by one.

~

IN SPITE OF the crowds, my protective daughters, and my stubbornness, I become known as a prized patient. My recovery is going well. I've been counting down each day since the skin graft surgery. Almost miraculously, on the tenth day, the pain in my legs is gone. *Gone! Just like that!* I'm on track to leave the burn unit before twenty days are up with near full range of motion in my arms. I'm somewhat mobile with the help of a walker. I hate to admit it, but those exercises worked wonders.

The glass embedded in my skin is still there; its too tedious a job to take it out. Countless shards of glass in my skin are meaningless compared to the burns, the debridement, and the grafts. There's just one thing left I have to overcome before I can go home.

Morphine.

From the ambulance ride until right now, my IV has provided a steady flow of the mother of all painkillers. Soon after I was admitted, they moved the IV out of my foot into a deep vein catheter in my groin. Not entirely comfortable, but at least it delivers the goods. I get a dose before everything: baths, daily debridement, dressing changes, even therapy. I have been so dependent on it that my daughters decide to give me a new name.

"Mom, you know you're, like, the mayor of Morphine Land," Brandi says.

"I'd like to resign from office," I tell her.

Jessica tells me, "As soon as you don't need the morphine pump anymore, we can let you go home."

It occurs to me that she might be my favorite nurse because she's always telling me the good news. Her words are magic to my ears. Getting out of the burn unit seems like a way out of the misery. I am determined to step down as mayor of Morphine Land. I'd rather be governor of Good Health.

On Friday, my eighteenth day in the burn unit, I try to do without morphine during one of my dressing changes. This is where a nurse unwraps all the tightly wound gauze and pulls off the snug gloves over my arms, replaces the healing ointment, and then wraps me up like a mummy again. Even though I feel like my pain is finally under control, a lot of force is used in a dressing change, pushing and pulling on swollen and tender skin. Just looking at my mangled skin is enough to make me sick.

We're in the middle of the dressing change. Suddenly, I feel hot and cold all at once. I'm about to pass out. I'm going to throw up. The nurse realizes I can't continue.

"What's wrong today, Cindi?" she asks.

"I didn't take any morphine this morning," I answer.

"Oh my God, you're doing this without pain meds? What are you thinking?" she asks with alarm.

"They said I can't go home until I'm off morphine. I thought I could do it."

"Cindi, we don't mean entirely. We just don't want you leaving while you're hooked up to a morphine pump, dependent on it all day. It's okay for you to take it before you go through a dressing! And once you're out of the hospital, we'll give you other pain medication."

I look over at Brandi. I feel like an idiot.

"Guess you just got reelected," Brandi says.

"To what?" my doctor asks as he walks in.

"Mayor of Morphine Land. That's what we call Mom. She's been trying a little too hard to kick the habit so she can get out of here."

"Well, then this should come as good news. We're imposing a term limit on you. You can go home tomorrow."

~

I CAN'T SLEEP. The anticipation is exciting—I'm going home two days early. Yet the prospect is scary. The burn unit, for all the torture and tears, has been a secure cocoon. The nurses and doctors have taken care of my wounds. My family and friends have not left me alone for a single moment. Except for the detectives, the press conference, and one news interview in my room, I have not had to face the outside world. I haven't even breathed outside air for eighteen days. *Eighteen days that have felt like eighteen months*. These thoughts roll around in my mind, and I can't turn them off.

Morning finally comes.

"How did you sleep, Cindi?" Jessica asks.

"Not very well."

"I'm going to miss you," she says.

"You made this experience as good as it could've been, Jessica."

"I know it's been so hard. But you did it! Ready for your last dressing change?"

My last one. I will have to come back every other day for a few months for new dressings, but this is my last one as a burn unit patient.

With a small dose of morphine, I tolerate it pretty well. I'm a little amazed. So many times, I did not think I would survive. And now I'm leaving. As a nurse comes in to wheel me out of the burn unit, I can't believe this day has finally come. All the anxiety I felt about leaving the night before is gone. I am ready for my own bed, my own space, my own life back again.

Angela pulls the car up to the ER entrance. Brandi and Shelli help lift me out of the wheelchair and gently boost me into the front passenger seat, which is pushed all the way back. My legs and arms are still heavily bandaged, so the extra leg room is welcome. Still wearing my gown, robe, and house shoes, I slowly get settled. At noon it's already hot and muggy outside, but Angela has cooled the car way down, just as I like it. I smell the fresh flowers the girls salvaged from my room. In the rearview mirror, I see boxes of cards, teddy bears, and balloons that had been brought to the hospital.

It doesn't take long for the girls to slip right back into our old comfortable routines.

"Mom, I'm starving. Do you mind if we stop?" Brandi asks.

"No, of course not," I answer. "A Coke and hamburger sound pretty good."

Angela pulls over at a place we've eaten together a million times—the nearby Sonic, the girls' choice for drive-in fast food.

This is the first burger I've seen in nearly three weeks. The drink feels so cold, sweet, and delicious as it rolls down my throat. But after two or three bites of the hamburger, I wrap it up and throw it away.

"Mom, are you okay?" Shelli asks.

"I'm fine." I lie out of love for my children. We've been through so much, and I want to protect them as well as I can. I start to realize how nervous I am about the ride home, but I don't want the girls to know. I sense that they're nervous, too, and I don't want to add to their anxiety.

Oklahoma City is a good-sized midwestern town of half a million people. This Sonic sits on one of the city's busier highways, just down the road from the hospital. As cars whiz by us, I've almost forgotten how big, loud, and powerful they are. I feel shaky inside, my palms sweat, and my heart pounds in my chest. I'm way out of my safety zone. These are not feelings I'm used to. I'm usually the one in control—driving when we're all together, leading the way, even setting the mood.

The girls probably thought since I'd been cooped up in the hospital for so long that I'd enjoy sitting in the car with the sun shining outside, the warm wind blowing in through the windows, and the familiarity of stopping at Sonic for a snack. But I am suffocating on my own fear. The traffic noise and the confinement of a car bring flashbacks of the morning we were attacked. I can't take it anymore.

"Let's go ahead and go now," I finally say.

My daughters don't hesitate. "No problem, Mom."

The girls haven't finished their sandwiches when Angela starts the car. I feel terrible for cutting their meal short, but I'm having a panic attack, and it's a long drive home to Duncan. The nurses had given me enough pain medication to make it home, but nothing is relaxing me. The girls try to make me as comfortable as possible during the hour-and-a-half-long drive, but if I can't find comfort with the daughters I love with all my heart, I can't find it anywhere.

The girls chatter a mile a minute, maybe even faster. They try their best to keep my mind off the trip home, but it's not working. I would have to be knocked out for that to happen at this point. I guess we could travel a different route, but going I-40 through El Reno would take us more than thirty miles out of the way and just delay the inevitable—eventually, with all the required return trips to the hospital for dressing changes for my burns, I'll have to get used to driving under the bridge where it happened.

I've traveled this route thousands of times, but I've never thought about the overpasses before or how many there are. I hardly even paid attention to them. When I'd looked up at them before, it was always just a cursory glance, never expecting to see anything more than a car. After all, they're just a means of crossing over a highway, or at least that's what they're supposed to be.

But today I'm looking for them. I want to know when they're coming up, and I want to know what is on them. As we approach them, one by one, they come so fast, and then they're gone even faster, until the next one. I'm keeping count in my head. *Fifteen, sixteen . . .*

Will I spend the rest of my days unable to enjoy the ride? Fixated on the overpasses ahead? Frightened to pass under them?

About the time we pass the highway sign for Newcastle, the girls quiet down. In a voice more unsteady than I plan on, I ask, "Is this the one?"

"No, it's the next one," Angela says, her voice barely audible.

As we come up to it, I try to keep my face perfectly inexpressive. It takes everything I have not to cry. If I start, no one in the car will be able to stop the flood of tears.

The bridge is empty. No cars, no people. There's no reason it should deserve anyone's attention, being so ordinary, narrow, and almost abandoned. I know I've never noticed it before. I'll bet most people who drive along the turnpike hardly even know it's there. I only learned the name of it after the attack: the bridge on Country Club Road. It runs through a sleepy valley with a few farmhouses and one pretty French châteaulike house on a pond. It deserves better memories than the ones we have.

For as long as I dreaded this moment, it's over, in an instant.

I feel the tightness in my throat loosen a little. Now that we're past the bridge, I can relax, which helps the pain medication do its job. I'm not comfortable, by any stretch, but I don't ache too much. My body is fifteen pounds lighter, pretty weak, stiff as a board, and I look like the hell I've been through. I'm really a pathetic sight, still in my robe and bandaged from head to toe. But as we drive into Duncan, it feels good to be home.

"Mom, now that we're almost home, we need to tell you something. Deron didn't want you to do this without him," Angela tells me.

"What do you mean?" I ask, curious.

"You know how you were supposed to get out of the hospital on Monday? Well he's been planning a big convoy to make sure you got home okay. He knew you'd be scared to get in a car again and go under an overpass."

My coworker Deron had arranged to have a CableOne truck waiting at each overpass on my way home. That would mean thirty drivers would be stationed under thirty bridges from the hospital to home. The plan was to wait for the car I was riding in to approach, and then they'd lead the way under the bridge, so that they'd go first every time, in a symbolic effort to protect me and my girls on our trip. Deron organized every last detail, including rallying some drivers from outside our Duncan office.

My girls knew about Deron's plan all along, but they couldn't very well persuade me to stay another day in the hospital. They knew I would jump at the first opportunity to leave, surprise procession or not.

"That is one of the sweetest things I've ever heard of," I say out loud. I turn to stare at the familiar scene out the car window, but the houses and street signs are blurred by my tears. I am touched that Deron and my colleagues thought to protect me, to make me feel like this kind of attack could never happen again to me or my children.

One act of violence spawning so many acts of kindness.

Turning onto my street, I see my neighbors working in their

yards, visiting with one another, and walking out to get their mail. They know I'm due home today, they just don't know what time. Angela pulls into my driveway, crowded with the kids' cars, and honks the horn. I take a deep breath before trying to get out of the car, which is something of an event. I move so slowly, one leg at a time, one arm at a time. Shelli brings the walker over to help me up. As soon as I get my balance, my neighbors realize it's me.

"Cindi!"

"Woo, Cindi!"

"Hey, Cindi!"

Clapping loudly and cheering from all directions, my neighbors give me a great welcome home. Leaning on my walker, I look around at all of them and start to cry like I've never cried before.

"Yeah, Cindi! We're so proud of you!"

The more they shout and clap, the more I cry, and I'm not the only one. Throughout the trip, it was not easy for the girls, but they kept their tears in check. They had been waiting almost three weeks for me to come home. This spontaneous welcome is just too much. All the tears they've held in for so long come pouring out.

It's just like a Hollywood movie where the whole neighborhood comes together and makes the place look like Mayberry. My whole life, I thought those scenes were nice, but just the stuff of fairy tales. But here it is, happening to me, and it feels wonderful. So maybe I don't look like a leading lady, but they treat me like the star of the show. I realize I never was in this fight alone. I had a whole community pulling for me, praying for me, and now cheering me on every step of the way. I take one hand off the walker and wave to thank them, since tears choke the words of gratitude I can't yell back.

"It's great you're home, Cindi! It's great to have you home!" Their shouts follow me up to the front stoop.

I know I'm where I need to be. The road ahead is long, but that doesn't matter to me now. I'm home, safe, and alive.

I stop again to wipe away my tears, slowly making my way toward the front door. I try to envision the inside of my house, just feet away, and wonder how much the girls were able to clean it up. When Jim and I had left for California, it was quite a mess, I'm embarrassed to say. We had ripped up the linoleum that had covered the floor for twenty years and replaced it with beautiful hardwood. Angela, Shelli, Chad, and I had finished it a few nights before the trip. But there are still half-walls with outdated pillars I really don't want to look at anymore. There is a list of projects to be done, including retiling the entry hall. I don't let the weight of all the work to be done worry me, because I know it will be months before I can even think about touching it.

Shelli opens the front door. There's Jim, smiling ear to ear. Steven's holding a video camera. Chad has Carson in his arms. Kennedy is jumping up and down.

And then I see the new tile under her little feet.

"Ohhhhhh!"

I start crying all over again. I stare in awe at the brand-new tile that starts at the front door and runs down the hallway. It's even better than having a red carpet rolled out for me.

"Oh my God!!!!!!!" I cry out.

Not only are all my kids and grandkids here, but they're standing in *my* house, which had been totally remodeled during my stay in the hospital. Shelli and Angela coordinated the massive effort, and it looks beautiful.

"Come look at the kitchen, look at the laundry room, look at your bedroom!" The girls can't wait to give me a tour. They recarpeted every room that needed it, thanks to one of my clients who owns a carpet store. She offered the carpet at cost and paid out of her pocket, as a gift, to lay it. They tore down the old columns, fixed the walls, laid new tile in the utility room, put new countertops and a new sink in the kitchen, replaced the hot water heater and refrigerator (which had both blown out while I was hospitalized), and moved a new armoire into the bedroom.

Angela explains they didn't do it alone. "Mom, Robin told us we had to get the house back in order. She basically said, 'Here's a check. Hire whomever you need to get it done. Whatever it takes.' "

Shelli chimes in. "Everyone who worked on your house knew it had to be done in time and just right for you."

I can't describe what it means to come home to this. I never dreamed it would all be done—even more than I imagined or hoped for! The parts of my former life I no longer wanted have been torn down to make way for the new. With each surprise, big and little, I cry and cry some more. *So much goodness.*

There are more cards, gifts, and food from family and friends. It almost feels like a birthday party, and indeed, we are celebrating life. I read one of the cards attached to a present from my friend Sharon.

" 'I'm so thankful you are home. I'm so thankful your face is beautiful again. I'm so thankful you have been so brave. I'm so thankful most of the physical hurt is over. I'm so thankful you are my friend. I'm most thankful you are alive! I love you.' "

The homecoming wipes me out. It's the most I've moved around in three weeks, and all I can do is collapse.

"Do you want to lie down in your bedroom, Mom?"

"I don't want to lie down. I've been in a bed for so long. I just want to be with all of you."

I spot my favorite rocker recliner, and it has never looked more inviting.

"I think I'll just sit in that chair right there."

The trip home and the emotion of the day have sapped my strength. My legs hurt so much because of scabbing from the skin grafts. I need to get off my feet.

Kennedy, who hasn't left my side, stands right next to my chair.

"Be careful, sweetheart," Angela warns.

All of Kennedy's attention is focused on me.

In a quiet voice, she asks timidly, "Can I rock with you?"

All at once, we hear a chorus of "No!" from the adults.

"Kennedy, you'll hurt Grandma's legs," Angela adds before I have the chance to say anything. Her nurse's instinct tells her how much pain I'm in, and how I don't need a squirmy four-year-old pushing against my aching body.

Looking at Kennedy's big brown eyes, and seeing the longing on her face just to have Grandma hold her again, I can't deny her, and I can't deny myself.

"Yes, of course you can. I've waited too long for this."

I lay a pillow across my right leg to protect it from being bumped. Kennedy knows not to sit on my lap. She climbs up gingerly on the other side of the pillow. I put one arm around her as we both scoot down in the chair into a comfy position and start to rock gently. I can smell her clean hair and feel her breath on my chest.

"I missed you, Grandma."

"I missed you, too, Tootie Girl. I'm so glad to be home."

I watch her eyelids get heavy, then shut. Before I know it, I join her in the sweetest sleep I've ever known.

7

Facing Fears

MY OWN BED has never felt so good. As I wake up and look around my room, I wonder if I'm still dreaming. The curtains are drawn closed, the sunlight behind them casting a red glow around the window. My room is cool and dark, just how I like it. Such a contrast to my sunny white hospital room where, although I seemed to sleep a fair amount, I never felt rested. This morning, I can't hear any movement in my house. The homecoming from the burn unit took everything out of all of us.

I shuffle over to the vanity and pull out the small bench. Slowly, I ease myself into a sitting position, in front of the mirror. Funny that it's called a vanity, because I've never thought of my routines here as an act of conceit. I always used to dread putting on my makeup, a chore I had to do every day of my working life.

Three weeks ago, the morning Jim and I were getting ready to leave for San Diego, I sat here thinking, *The only thing left for me to do this morning is put on my makeup. I'm going on vacation, and I still have to put on this stupid stuff.*

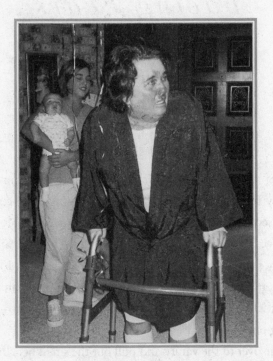

Taking it one step at a time

"Putting my face on" had become such a bore for me. *How mundane is my life?* I'd asked the mirror. *I get up, shower, brush my teeth, put on makeup, get dressed, and go to work.* The routine was set in stone. *I hate every second of this,* I'd tell myself every morning. What a way to wake up to a new day.

Applying makeup became the focal point of my unhappiness. A fingerful of base coat, a dusting of powder, a brush of eye shadow, a layer of mascara, a coat of lipstick, all layered on top of a frustrated face. Once I even tried to change the order. *This will bring something new to my life!* I put the lipstick on before the eye shadow, and—you guessed it—no difference. No matter how hard I tried to make the routine new and different, I couldn't find a way out of the rut.

Well, I had found a way out all right! One quick acid attack later was all it took to shatter the dreaded routine. As I sit here again, looking at my reflection, at the tortured skin on my face and body, I can't believe I actually fretted over something so insignificant, that I wasted time in my life making mountains out of molehills.

I thought I needed a new routine, but I hadn't counted on it being part pain medication, part dead skin removal, and part bandage changes for eight more weeks. I don't imagine that in the months to come countless trips back and forth to the hospital will be a joyride either. Not to mention the strength it will take to drive past the crime scene every single time.

I swore that I'd never put myself through the misery of skin grafts again. But my doctor tells me reconstructive surgeries—as many as a dozen of them—are a matter of necessity, not vanity. Every three months or so for the next three years, I can expect to endure an involved, four-hour-long surgery. The first and second surgeries will rebuild my upper and lower eyelids so that my tear ducts aren't exposed. My eyes are terribly dry and my vision is so blurred,

I can't read anything. With burns in the corners of my eyes, I can't close them completely. Grafting in those precarious areas has to be done by a plastic surgeon.

What's true for all of my surgeries is that fixing one thing will bring on pain somewhere else. For example, the skin for my eyelids, upper lip, and chin will come from my stomach. Stomach skin is supposed to be the closest match in color and texture to facial skin. Since he'll already be making a five-inch incision, I wonder with some amusement whether the surgeon can give me a mini tummy tuck at the same time.

The burns around my mouth tighten more and more each day, so much that they nearly pull one nostril closed. I will have to wear a device to widen my mouth—a metal screw to twist a quarter of a turn every day to pull it back open. Doctors also tell me that I should count on at least two surgeries to widen my mouth. I can open it just enough to get in teaspoon-sized bites of food, but the metal mouthpiece won't be enough. I'll need a surgeon's blade before I can eat a regular meal or even fully smile again.

As for makeup, the doctor says I probably won't put it on for another two years.

But that doesn't matter, because makeup has nothing to do with how I see myself anymore. When I had my old face, the one that wasn't burned, the one that could smile fully, I saw a middle-aged, somewhat overweight woman whose full potential had not been realized. The acid burned so much of that away, literally and figuratively.

Now my reflection shows me something else. First, it tells me that I am alive. What a privilege it is to get to sit in my room, at my vanity, and look in my mirror!

Second, I don't see a face anymore; I see a person. I see a survivor, a loving mother, a fantastic grandmother. No matter how many surgeries I go through, I wonder if I'll always look like a burn victim. But I'm not worried about that. I've separated my self-image from my body image. Phil has talked to me about it before, how the line between those two things has blurred in our society, how people value what they look like on the outside more than what they are on the inside. But I know the difference. Even though my body has changed dramatically, the things that make up my self-image— God-given talents, my passions, my soul—are still there. I'm at peace with who I am. I know I'm the same person I was before this happened. I can't let my self-esteem deflate because of something that really is only skin-deep.

My broken surface tells an incomplete story; all the relationships that matter most to me are intact, if not more meaningful than ever. And life is all about those connections, not one reflection in the mirror. The eyes of my children and their children are my mirrors now. This realization is a defining moment in my recovery, a critical part of the second crossroads, a choice I make to help myself heal. Phil often talks about episodes that are like epiphanies. Defining moments. Moments that mark a permanent change or turnaround, from an undesirable condition toward a favorable goal.

I'M GOING TO need this new sense of self before going out to face the rest of the world, the world beyond my home and family.

I start with a relatively safe place—facing my coworkers at CableOne. I'm not in great shape, but I want to see my colleagues and thank them in person for their support. Between my physical condition and the weather, getting around is no picnic. The summer heat is hitting record highs with several straight days of temperatures over

100 degrees. My doctor has insisted I stay out of the sun. I'm hot-natured anyway and would be uncomfortable even if I didn't have so many injuries to contend with.

Brandi comes to pick me up in her little car. Her air-conditioning is on the fritz.

"The air coming out of here is just not cold. I feel like I'm in a convection oven the way it's blowing on me," I tell her.

"I know. It's been like this all summer," she explains.

By the time we pull up to my office, I'm sweating. Brandi pulls my walker around and holds a black umbrella over my head to keep the sun off my skin.

I'm still covered in bandages and haven't had my hair washed in a few days, but I'm determined to see my friends. Slowly, I walk down the back hall, with Brandi just a few steps ahead.

She announces, "Guess who's here?"

Immediately all the familiar faces gather around me. Everyone starts talking at once.

"Cindi! You look great!"

"We're so happy to have you back."

"Oh my God! I can't believe you're standing here."

Their loving words mean so much to me. I know I've made the right choice to come see everyone. They lift my spirits, give me more strength, and reinforce my courage. But after a few minutes, I realize I'm exhausted from standing.

"I've got to sit down." Before I know it, there is a chair right under me. My friends are eager to take care of me in any way they can.

"When are you coming back for good?" someone asks.

"As soon as I can get my hair done!" I joke, and everyone laughs with me. "Don't you dare give my job away."

After the visit, Brandi and I go next door to the beauty salon where Shelli works. She had offered to wash my hair and do it up whenever I was ready. I try to lie down and tilt my head back over the sink, but the incisions around my neck pull too much. Shelli improvises with the hair-washing, gives me a much-needed trim, and makes me feel like a million bucks.

A week later, on the way back from a dressing change, I make a last-minute decision to drop in on some valued clients, Buzz and Lee Ann Carter.

"Do you mind pulling over here?" I ask Jim, who has continued to be a faithful friend and chauffeur.

"No problem."

I haven't seen any of my clients since the attack. I don't want them to feel pity for me or to feel uncomfortable because of the way I look. I have agonized about the day I'd face business associates outside my home office. Just this week I'd received a card from family friends. "I saw your interview on TV. You did very well. Maybe you should think about going into radio or TV when you recover. Maybe into selling ads. Our thoughts and prayers are with you." Their soft sarcasm made me feel relaxed about two things: they'd already seen me on the news, and they still wanted to joke around with their friend.

I am not, by nature, a worrier. "Don't borrow worries," I tell my girls when they fret about something that may or may not come to pass. Most things we waste energy worrying about don't actually happen. And the little things we worry about likely mean the big pictures of our lives are pretty good just as they are.

As much as I try to not worry, however, I have to admit to some fears. I believe there is a difference. My fears are rooted in the searing pain brought by the tiny stick of a needle and the deep cut of a surgeon's blade. My fears come from knowing human nature and how it reacts to disfiguring scars. My fears are based on my experience of being attacked in the dark of night while driving on a road I'd traveled hundreds of times, a road I'd always believed was safe.

Facing my clients again is another fear. But as we near Buzz and Lee Ann's place, I feel emboldened.

"Why do you want to see them first?" Jim asks.

"The very first bouquet of flowers I got in the hospital came from Buzz and Lee Ann."

The Carters own a small car lot and have had an advertising account with me for years. We'd become friends through business.

"Oh my gosh . . ." Buzz says out loud as I walk into his office.

"I'm so, so glad to see you, Cindi Broaddus." Buzz gives me a big but gentle bear hug. Lee Ann bursts through the door, a smile on her face and tears in her eyes. She, too, hugs me carefully.

"Thank God you're alive."

"Sit down and visit with us awhile."

I can't describe how good it feels to know the Carters are truly thrilled to see me. It reminds me that I enjoy selling for CableOne not because of money, but because of the relationships I develop with my clients. And I knew whom to go to first. I knew they would welcome me with open arms and hearts no matter what. That short visit with Buzz and Lee Ann boosted my confidence and got me ready to face the rest of my clients as well as the working world.

"Buzz and Lee Ann are good people," Jim says as we drive away. "They almost make up for all the other stupidly cruel people who stare. I hate that geeky stare. You know, the one where their lower jaw drops below their chin?" Jim opens his mouth wide, doing his best impression.

"Jim, that's not humanly possible." I don't mind Jim's tendency to exaggerate; his version of life is always so colorful. And in this case his defense of me is very sweet.

"I know, but you understand what I'm saying. If I could do anything to change an aspect of what happened, it would be the way people stare at you. It offends me. When I look at you, I don't see your scars, Cindi. I see you. It hurts me because I know how pretty you were." Jim shifts his gaze from out the window to me. "How pretty you still are."

The way I feel about Buzz, Lee Ann, and Jim brings another lesson home to me. You never know how what you do now will affect someone's life later. I guess it's kind of like the adage "What goes around comes around." I've always tried to show my friends kindness and respect, and they are giving it right back to me at a time when I need it as much as I need air to breathe. But I don't apply "what comes around" to mean destructive, and random, behavior. Otherwise, what did I do to deserve being attacked? I choose to believe it's goodness—intentional goodness—that makes the rounds.

A few days after visiting the Carters, I decide to make a big trip to the mall with Brandi. I have just enough energy for one outing a day. I'm still moving slowly, and under the bandages, my legs are terribly scabbed and itchy. We make our way into a jewelry store owned by a client I'd always liked a lot.

"Hiiiii, Terry," I say tentatively as I approach him at the counter.

"Cindi! How are you? I'm so glad to see you!" Terry's eyes show genuine enthusiasm.

"I'm trying to get in a groove here, to get used to being out in public. Hope you don't mind me dropping in."

"Are you kidding? I'm so grateful you've come." Terry makes me feel right at home, as if I look like the same person he's always known.

As we say our good-byes, Brandi and I turn to walk out. Terry is still standing at the counter when he yells in our direction.

"Cindi, you dropped something!"

I look down and almost die from embarrassment.

"Thanks, Terry. Just a huge piece of my leg!" Some of my bandaging had loosened from walking around so much, making way for an enormous scab to fall off my leg and land squarely in the middle of my client's jewelry shop. It must be the size of a silver dollar and is truly disgusting. *Nice calling card. Very nice. Oh my God.* Here I am trying to gain confidence about public appearances, and I have to go and leave body parts behind. I bend down and pick it up on the way out. *Where do I put this thing, in my purse?*

"My God, Mother, that was absolutely mortifying," Brandi whispers as we leave the store. It was mortifying, but more than that, I'm starting to find the embarrassment downright hilarious.

"Could you imagine if *he'd* picked it up after I left? He would've had to vacuum the floor after me!" Brandi and I start to giggle and don't stop all the way home. I love that you just can't plan for moments like that. Life makes you laugh when you least expect it. And even though the experience was a real humiliation, I know I can face anyone and handle just about anything.

The next week, three weeks after being released from the hospital, I return to work part-time. The plan is not to see the public right away, but to work the phones, making sure to call ahead and give my clients, new and old, a warning about my appearance. This is not going to be easy, but I have two choices: sit at home for months doing nothing or motivate myself to go back. A voice inside me tells me I have to get up and go. It's time to get back in the saddle. Mentally, I know I can do it. And my coworkers have let me know they are here to hold me up if I fall.

"When can you come out here?" one of my clients asks.

"Let me forewarn you. I've been in a major accident and I don't look so hot," I tell him.

"I don't care what anyone looks like. You come on in. We'd be honored to see you."

I had no idea my clients would be so welcoming. Brandi is about to be very busy driving me all over town. My car was totaled in the attack, and I'm not able to drive anyway. Even with a short workday and my own chauffeur, I'm physically drained by week's end.

On the way home, the heat in Brandi's car is stifling. I can't take it anymore.

"Beebop, I'm putting a whole new air conditioner in this car."

"Aw, Mom, I'm so glad you're back." She smiles at me for spoiling her.

"Yeah, me too." I want to be strong for Brandi, who has extended her leave from work this summer to help me. But in truth, I wonder if I haven't gone back too soon. My energy is zapped, and I feel like I've been in a train wreck.

"Do you think we could go by Angela's first?" I ask.

Angela's back door is open as I shuffle into the kitchen. I must look like the walking wounded, because when Angela sees me, she immediately asks if I'm okay.

"You don't look very good. In fact, you look kind of bad."

Kennedy, who's in the den watching TV, whips her head around, jumps up, runs into the kitchen, and wraps her arms around my leg. It hurts, but not too much.

Still attached to my leg, Kennedy looks up at her mother. "What did you say?"

Angela tries to explain. "I just asked Grandma if she was okay because she doesn't look very good."

Kennedy's tone is defiant. "Don't you ever say that again, Mom. My grandma could never look bad to me."

Angela and I look at each other. I know Angela is concerned about me and doesn't mean to insult me.

Suddenly I'm beaming.

I realize how much I need Kennedy today. I'm not sure why I went to Angela's instead of going straight home to collapse in my own bed. Angela had been cautioning me to go easy, to make sure to take breaks, to not do too much. I've obviously overdone it, and I'm feeling pretty down. Kennedy not only turns my mood around, she marks another defining moment. I will never forget my grandchild taking hold of my leg and defending me to the person she loves most in the world.

As I leave their house, my heart feels lighter. Kennedy came to her grandma's rescue today.

I'm a different person as a grandmother. The more I realize that, the more I understand that even an acid attack can't take anything away from me. Like any parent or grandparent, I think my kids are flawless, and though I know I am flawed, scarred, and far from perfect, they treat me like I am all that they need.

But loving my children and grandchildren so deeply comes with a price.

⌒

I LOVE SURPRISES, and I know Shelli will welcome this one.

It's a warm and windy Wednesday afternoon in July, and she has an appointment with her obstetrician.

I call Shelli's doctor ahead of time and tell him what I'm planning. "I don't want Shelli to know I'm coming, but I know she'll want me there."

The waiting room is filled with big-bellied women, mothers and children, and skinny girls starting to show slight pooches, just like Shelli. I wave at the receptionist with a childlike excitement, and she walks me back to the examination room where Shelli is lying down.

"Shubug?" I say as I pull back the curtain around her bed.

"Mom! I didn't know you were coming! Look!" Shelli has a smile on her face that seems wider than her tiny belly. She looks down at her barely rounded stomach.

"Listen to this, Mom."

Whoosh, whoosh, whoosh. Whoosh, whoosh, whoosh.

"Ohhhhhhhh," I moan as tears fill my eyes. I'd heard the sound myself, three times before—for Angela, Shelli, and Brandi. "My grandbaby's heartbeat." I can barely get the words out. Shelli reaches for my hand.

"I love you, Mom."

Moments like this, holding my daughter's hand, and listening to her child's heartbeat, lift my spirits and make everything else fade away.

Initially, I thought I was there to surprise Shelli. I was so focused on supporting her that it hadn't occurred to me what I would do for myself. It doesn't matter that my arms hurt and my face is swollen and my mouth can't move much. I am still Shelli's mom, and she's about to find out firsthand why I love her so much.

One month later, Angela calls me from the obstetrician's office. She's gone with Shelli this time for her routine checkup.

"Mom, they can't hear a heartbeat. Shelli's waiting on another sonogram to check again."

I can only imagine the disappointment Shelli is feeling. She wants so badly to be pregnant, to feel her belly grow, to bring a baby into the world. She beamed just knowing she was going to be a *mom*. She had woken up today, so excited to hear the miraculous *whoosh whoosh whoosh* sound, only to hear silence.

I am not prepared for so much sadness so soon after my trauma. I drop what I'm doing, call Jim, and arrive at the hospital just before Shelli undergoes a D & C, the medical procedure to remove an unsuccessful pregnancy.

I almost feel like I'm not here, like I'm watching all the sorrow and disappointment and tears from afar. Honest to Pete, I'm not sure I can handle much more. I feel myself closing off. To see my child so unhappy, knowing how badly she wants a child, brings me to my saturation point. *I'm so sad, I don't think I can be any sadder.* It's killing me to see Shelli so upset. I look at Angela with Carson, her adorable four-month-old boy, and think how Shelli has just lost hers, and how randomly the pendulum of life and death seems to swing.

Carson and I bonded early in life. I was in the room when he was born.

Having raised three daughters, I wasn't certain how I'd relate to a grandson. I'd never really been around little boys. I was excited to have a grandson, but I was also a bit concerned. I knew I'd been a good grandma to Kennedy, and I was hoping I was up to the new task.

Minutes after Carson was born, our celebratory mood turned serious—he had trouble breathing. The pediatrician was called in to evaluate Carson's lungs, and I could tell the medical staff was worried about him.

I'll never forget walking quietly over to Carson's crib and bending over him. His eyes were wide open, unusual for the newly born, and they stared right at me. I decided it was a good time for a talk.

"Carson," I whispered so as not to startle him, "we've been waiting a long time for you, and your mommy is so excited you're here. But you're not breathing very well, and that makes your mommy sad. So you need to step up, be a big, strong boy, and breathe better. Okay?"

A NICU nurse stood next to me. "Keep up exactly what you're doing," she said. "His vital signs are picking up. He's already doing better."

From that moment on, I never worried again about whether Carson and I would have a special connection.

I know Carson can't possibly remember the first moments we shared together, but these days he's the one trying to take care of me. The only reason he points out my scars is because he's worried about my comfort.

"Does it huwt, Manya?" he often asks.

"No, little man, nothing hurts when you're around."

~~~

IN THE DAYS AFTER Shelli's surgery, Angela takes me shopping as a distraction. We head out to the town of Chickasha, forty-five minutes from Duncan. Now that I'm back at work, most people in Duncan are used to seeing me out and about, and they don't seem to gawk as much anymore. But walking into a new town to face new curious eyes is more uncomfortable than I expect.

We go into the store in a hurry, buy something, and then head back down the main aisle. I'm just ahead of Angela, but not too far away to hear her sigh. This is no ordinary, oh-I'm-kind-of-tired sigh. It is loud and deliberate, one that tells me something is annoying her deeply. I turn to see what it is. One of the store clerks has stopped dead in her tracks and is staring at me like I'm an alien who has just landed in aisle five. Her expression is exactly the one Jim despises so much. Angela glares at the clerk, and the clerk scowls back at Angela.

The two of them are literally having a stare-down in the middle of the store. Fists clenched, Angela finally breaks the standoff and stomps out ahead of me to the parking lot.

"Mom, that makes me so mad," she says as we get into the car. "People are so pathetic. They should have more class than to stand there staring like you were something from a horror movie! What if it were *her* mom? People just don't think."

"It doesn't bother me when people take a second look. I think that's a natural reaction. I do it myself," I tell Angela.

I wouldn't say I welcome chances to talk with strangers about my injuries, but I'd much rather they ask than stare. I prefer to have the opportunity to explain that, yes, something terrible happened, but that I've seen so many acts of kindness because of one act of violence.

"Remember the little girl in the Wal-Mart last week?" I say. "Her, I did not mind."

I had been looking at digital cameras, all the while feeling a pair of probing eyes on me. I looked around and didn't see anyone, until something down below, about three feet tall, caught my attention. A little blond girl was staring up at me. She was so entranced that she hadn't noticed her mother had already moved a few paces ahead.

Finally, the four-year-old mustered the courage to ask, "What happened to youw awm?" Her voice was tender and true, and her inability to pronounce her *r*'s made me want to scoop her up and give her a hug. *This tiny thing is braver than most adults.*

I leaned down and smiled. "Well, thank you for asking. I'll tell you what happened to it," I said as if I were beginning a bedtime story. Her eyes grew wide.

"Okay," she replied shyly. I explained that I'd been burned by someone but that so many wonderful people were helping me get better.

"Sometimes people do mean things. We don't know why, but guess what? Now someone is doing something nice by trying to fix it."

The little girl seemed satisfied with the story and went on her way with her mom. Some parents fear embarrassment when their children ask strangers questions, but hers didn't bother me. The two of us treated each other with respect. She was inquisitive about something, and I fulfilled her curiosity.

I try to teach children and adults alike that you can live through something as awful as this, you can end up looking quite different, but it doesn't have to leave such terrible scars on the inside. If anyone were to check my sense of humor, my happiness, or my heart, they'd see those things are unscathed.

The little girl knew all of that naturally, but the clerk Angela stared down had no clue.

"When people stop, stare, and gawk, that bothers the heck out of me. But it obviously affects you more. I thought you were going to deck that lady." I laugh.

"I should have. Maybe when people stared at the big purple shiner on her left eye, she'd get a taste of what it feels like."

Angela's eyes are welling up with tears.

I give her shoulder a squeeze. "Angela, it's okay. I'm okay. Don't worry about that stupid lady." It hurts me to see her so upset.

"It's not that, Mom. I can't believe how well you're handling all of this. It takes guts, Mom."

As protective as my girls are, they can't shield me from public opinion. As funny as it may sound, after years of struggling *before* the attack, I'm finally comfortable with who I am. Being a burn survivor, and looking like one, is now who I am, and I'm fine with that. I'm not as comfortable about the fact that so many people are just plain rude and unthinking. I never realized it when my appearance was normal. What I want to say to them is, *Imagine if I were someone who was upset by your stare. Imagine how truly terrible my life would be.*

~

"WHILE YOU WERE ASLEEP, the whole world was blown away," a nurse tells me when I wake from the first surgery since leaving the burn unit. The morning of September 11, 2001, I am in the Zale Lipshay Hospital in Dallas under the supervision of Dr. Rod Rohrich, a well-known plastic surgeon.

This surgery will start the process of healing the eyes I can't shut and the lips I can't close. As I am wheeled into the recovery room, everyone thinks I'm out cold, but I have a slight level of consciousness. In my room, where my family is waiting for me, I can only just see Robin. She's turned toward the television set. The volume is barely audible, but it sounds like CNN.

"Two towers of the World Trade Center are no more . . ." the anchorman drones.

Without opening my eyes, I moan back. "You've got to turn that off. I can't take it."

"Oh my God!" Robin is shocked to hear my voice. "She can hear it. Turn it off! Turn it off." Someone flips the switch, and the room falls silent.

I believe I am numb to everything outside my own misery. But hearing that people are dying by the thousands is devastating. *Did the nurse really say we were bombed, or is the anesthesia doing strange things to my mind?*

A few hours later, I wake up and the horror of the day starts to sink in. From my hospital room window, I can see the Dallas Trade Center piercing the clouds, and yet New York's World Trade Center is nothing but a burning ruin. It is harrowing to make the connection, and I am profoundly saddened. Mixed with visions of the horror people suffered are memories of my morning of terror, and I feel traumatized all over again. Yet I keep thinking how much luckier I am than the poor souls who did not survive. All I can do is pray for them and their families. And the more I think about them, the less I sleep. September 11 is one of those days everyone will remember exactly where they were when it happened. I'll never forget where I am—at my own personal Ground Zero.

AFTER EACH SURGERY in Dallas, Robin and Phil open their home to me for my recuperation. My sister waits on me hand and foot, making me as comfortable as possible. The days seem to pass quickly; there's always someone to keep me company. But when the lights are turned off, and everyone else is sleeping, I'm on my own.

Ever since the attack, there are times I struggle to fall asleep. The metal contraption meant to widen my mouth is impossible to wear during the day because of all my drooling, so I try to put it in at night. Every few seconds, I make a loud slurping sound, trying to contain my own saliva. I tried stuffing cotton balls in between the metal bars and my mouth, but they were instantly soaked. Once I do manage to fall asleep, the next morning my lips are just as dry as my pillow is wet.

I also have to contend with two terrible fears concerning my mouth. If, for some reason, I have to throw up, I'm afraid that the opening is so small that I'll choke on my own vomit. The other fear is that I'll get a toothache, and the dentist won't be able to do anything about the pain.

I can't get my mind to focus on anything relaxing. When I close my eyes, I can't see anything, so I'm not even able to picture the ocean or a meadow or sunshine. I have this unnerving feeling that my world will explode, from blackness to fire. I am barraged with flashbacks, and my brain pulses at the same quick pace of my heart. There is no sound, not even screams from me or Jim, just a void that offers no tranquillity. I feel an involuntary desperate need to replay the night we were attacked over and over again in my mind, so that I can remember exactly what happened. If I could see, hear, or smell it coming, then my memory could rewrite this terrible history, and I could foresee it, avoid it, and leave it behind. I have to wake up from it in order to find the peace I need to fall asleep.

Robin notices my sleeplessness. Worried that it will affect my ability to heal, she voices her concern to Phil, who immediately comes up with a solution. The next morning, Phil asks if I'm willing to try an approach to sleep therapy that might be unusual for me, a little out of my comfort zone.

"You need to trust me on this, Cindi."

"I do," I say, but truly, I have no idea what that will entail.

A colleague of Phil's comes over to the house and pulls out a set of bongo drums. He starts to play a soft, rhythmic beat. Listening to a man playing on drums is way outside my idea of normal, but I am desperate.

In a calm voice, he asks me to close my eyes and think of something peaceful. My mind drifts off to memories of a trip I recently took to Cancún with my sister Jamie. I'd never seen sand so white or water so blue.

"Relax," says the voice in between drumbeats as he guides me through a meditation.

"Let your body and mind see you are pain-free . . . *thump, thump,*

thump . . . This is not yours to worry about . . . *thump, thump, thump* . . . Law enforcement will take care of him . . . *thump, thump, thump* . . . You take care of you."

I stifle a laugh. This is not what I expect. Phil calls this sleep imagery, but the only image I can conjure is Desi Arnaz.

But I keep my eyes closed, and before I know it, I am drawn into the soothing voice and hypnotic rhythm.

I don't know how long I'm sitting here, but when he tells me the session is over and it's time to wake up, I don't want to. The bongo drums work. He gives me a tape of our session.

When I return to Duncan, the fears I have about my mouth actually happen. One of my medications makes me sick to my stomach, and I throw up. Somehow, small opening and all, the body compensates for almost any shortcoming, and I don't choke to death. I also crack a tooth biting down on a cracker, and only God and the dentist know how he gets in there to repair it, but he does.

Worst of all, I have to spend my nights alone, something I haven't done in months.

But I have the drums, and they work a miracle each night, lulling me to a deep, restful sleep.

"Standing Together" with my sister Robin

Quality time with Phil

8

In the Spotlight

One week after my surgery in Dallas, Phil calls me at home.

"I'm doing another workshop in Dallas in a few days. We're going to talk about pulling together in the aftermath of terrorism. Will you help me?"

Prior to the attack on my life, I'd volunteered for Phil's "Relationship Rescue" seminars in big cities like Atlanta and Chicago. I essentially filled the role of goodwill ambassador: helping people participate in workshops, ushering them to their seats, providing anything they needed. It was a very behind-the-scenes position, and I loved it. The trips never cost me a thing, and the staff was treated like royalty. Traveling around the country was a whole new experience for me, and watching the way people responded to Phil's wisdom during the workshops was even more enlightening. Phil would run marathon five-hour sessions, broken up for breathers every hour or so. And the most incredible thing would happen when we'd come back into the room after a break—Phil would not have lost one person. There was never an empty seat, a slip out to the hallway for a cell phone call, a rush to get to another appointment. Even after the five hours were over, people would stand in line for a half-mile to get Phil's autograph or photograph, linger to buy his latest book, and talk with him until he was hoarse. No one worked harder than Phil, and I was proud to be on his team.

I had gone on eight trips with Phil throughout the year 2000, and I had learned a lot from those workshops. Phil's philosophy about relationships always struck me for its simplicity. When he talked, nothing sounded complicated, although I knew how complex my relationships, especially my marriage, had seemed. "Would you rather be right or happy?" was one of Phil's best questions. "You have ten pivotal people in your life who have made a difference" . . . "Pay attention to those defining moments" . . . "It takes two to make—and break—those marriage vows" . . . "Do you know we often treat our friends better than our spouses?"

Now that I've been through a life-changing experience, everything he says makes perfect sense to me. But even before my ordeal, I thought about how I'd tried to live the way he was describing, to teach those ways to my kids, and to overcome mistakes I'd made in the past. My ex-husband might've broken our marriage vows, but I was as much a part of breaking the communication, respect, and friendship. If I were to ever marry again, I would conduct myself differently. During one workshop, I played an old scene between David and me in my head.

I'd be tired after a long day with the girls. He'd be tired after a long day at work.

"What's for supper?" would be his first question.

"Why don't *you* fix it?" would be my annoyed answer.

Given a second chance, I'd work harder at asking how his day went, allowing him to vent, and then work on pleasing him—and, ideally, he would reciprocate. I try this formula with my daughters and grandchildren, although they are arguably more lovable. I hug them when they arrive, ask what's going on in their lives, or, at the very least, show them a smile when they walk into the room.

I never got tired of going to Phil's workshops, but the attack in June made the chance of another one any time soon quite remote. When Phil asks me, even though it has only been nine days since my surgery in Dallas and three months since the attack, I welcome the chance to help out again.

Most of all, I realize, facing the public—looking the way I do—will make for good therapy.

"Are you sure you're up to it, Cindi?" Robin asks out of real concern.

"I'm there," I tell her, without a doubt in my mind. If I'm able to hobble around my house, surely I can hobble around Phil's workshop. It just doesn't occur to me to say no.

I still have that run-over-by-a-Mack-truck look, thanks to the swollen skin stitched painfully together. I can't close my right eye, and the surgery left it bruised purple and black as if I'd been punched. The skin around my mouth is discolored from the grafts and keloid scarring. The edges of my mouth have two large scabby sores on either corner as a result of my surgeon's attempt to widen it. I suppose I look worse to people who don't know, because I can see the way their eyes either quickly divert to the floor or stare without blinking. I expect to see a lot of those folks down in Dallas.

⁓

BEFORE GOING TO THE seminar, I stop by a bank in Lawton, a town thirty miles west of Duncan. The front clerk is clearly surprised, if not shocked, by what she sees. Right away, she brings up a safe topic—the weather.

"It's chilly out there for a sunny day," she says.

"The cold really hurts my scars," I say.

"How did you get them?" she asks.

Instead of going around the subject, she goes right for the truth, and I'm happy to tell her. For half an hour, we stand at the counter together, as she asks questions and I tell her what happened. Finally, I notice she's crying.

"I'm so sorry to upset you," I say.

"No, you're not upsetting me. I can't help it. It's so horrible, but the way you tell it, it's the most beautiful story. That your heart responds to such random violence with hope and kindness. I applaud your courage."

"Well, I appreciate that. But it's not mine. It's the courage of all the people who have stood by my side through all of this."

Here we are—perfect strangers, sharing a personal story, even tears, in the middle of an out-of-town bank. Little does she know that she has just become another person to help steel my nerve. More than ever, I am ready for Dallas.

THIS WORKSHOP IS particularly important because of September 11. Even though New Yorkers have been attacked, everyone in America is hurting, and we all feel vulnerable. The media had already looked to Phil for advice on healing the country's broken heart. Even Oprah arranged a last-minute taping for the First Lady and him to share their opinions on helping young children handle the tragedy. *Entertainment Tonight* will be covering the event. More than a "Relationship Rescue," Americans need a way to soothe their souls. Phil has been preparing just the right things to say to the eight thousand people gathered, some of whom are even a little fearful that such a big assembly could be the next terrorist target.

Brandi, Kendra, and I pass out blue button-down shirts that I've made for the staff. I think they look pretty cool, with DR. PHIL stitched on the front, and WE STAND TOGETHER embroidered on the back.

The program is about to begin, so we stay in a dressing area until it's time. All the workers I know from previous workshops stop by to say hello and give me a warm welcome back. Backstage, there is a palpable excitement that feels different from any of our other assignments.

"Hey, guys!" Phil pokes his head into the room. "I didn't realize it, but the events center has its own crew to work the crowd, so you're free to fill up the front row and just enjoy the program." Phil is so nonchalant about it that I don't think anything of the change, except that I am excited to have a place to sit for the next few hours.

"And that's where these terrorists screwed up, ladies and gentlemen, because they underestimated America. They attacked one of us, they attacked all of us, and it galvanized America! It got America back on its feet and said, you will not attack one of us without attacking every one of us!" Phil is on fire, like a veteran preacher in the pulpit of his home church. The crowd is revving him up, too, jumping to their feet and even waving their arms. I think they might just shout "Amen!"

"September 11 made people pull back together. It starts at your house and in your heart. We've got to have pride in who we are. We've got to have pride in America. And we have to show the world, and anybody who wants to attack us, that they're dealing with people of passion and they are dealing with people of courage! We bond as a country, we bond as a community. We find meaning in our suffering and we say to the world, I will not surrender my lifestyle. I will not surrender my lifestyle! You will not take our country, you will not take our life, you will not take our spirit, because you know what world, WE—ARE—STILL—HERE!!"

I nod my head with such conviction, I fear my neck will hurt the next day. *I am still here, and I'm going to make the best of it.* Phil is doing a masterful job of making our national tragedy feel very personal.

"Turn to someone near you and thank them for what they've done," Phil instructs the crowd. The rumble of well-wishers fills the massive center, and it reminds me of the way passing the peace during church service breaks the silence as well as the ice between strangers gathered for a common purpose. The feel-good exercise has the room buzzing.

I spot a handsome young black police officer standing stoically by the stage. His arms are crossed and his stare is sharp like a radar. For some reason, I feel drawn to him. While most people shake the hand of the person next to them, I rise from my seat and walk over to the officer.

"Thank you for the job you do every day," I say, and I mean it. To me, police officers are some of the most underappreciated people. Hundreds of thousands of them put their lives on the line every day, and all you hear about are the handful of bad cops. I can't help but think about the way law enforcement has been looking out for me ever since the attack on my life.

The officer sizes me up, wondering why I single him out. With a guarded expression, he quietly says, "Thank you." I don't mind that he is so standoffish; that's exactly what he's trained to be in enormous crowds like this one, especially in the shadow of the most devastating attack ever on American soil.

When we sit back down, I look over at Robin, who is just across the aisle from me, and we smile at each other, the way sisters can when they feel the special bond of the past and don't have to say a word to enjoy the present together. She and Phil love each other so much, and I know she's proud of his performance. I, too, think his speech is brilliant. I feel that his words are not only making sense to the crowd, but also deeply touching them. *This audience will walk out at the end of the day changed people.*

"If someone in your school, your church, or your community is affected, do you think it's any less important to someone who is hurt in your community, than someone hurt in a national crisis? We've got to be givers again. We've got to give in our own world, in our own lives. You may know our life has been hit in that way, because my wife's sister, Cindi Broaddus, was unbelievably attacked in the most despicable way I can imagine."

Robin and I turn toward each other again, and this time our eyes well up. When Phil starts to talk about the attack on my life, in front of all those people, I lift my bandaged arms to my scarred mouth, and try to hold back my tears, but it's hopeless.

"Let me tell you about Cindi. She's just like all the rest of us, just kind of a grassroots old girl. Lives in Duncan, Oklahoma. Incredible, incredible woman. Single mother of three children. Put all three of them through college. Worked three jobs. Had a day job from eight to five, but got up every morning and threw a paper route. And when she got off work, she mowed yards. A proud, strong, independent woman, who works for a great company—CableOne in Oklahoma. She worked so hard. She was headed to the airport, to take her trip of a lifetime."

Uh-oh, I think. My heart drops to my stomach. *I'm going to kill him. He's going to make me stand up and wave to the audience or something.*

"On the way to the airport, she went under an overpass and some moron threw a vat of acid that hit her windshield, and it came through, and put third-degree burns all over her body. An absolute senseless act of violence. I'll share it with you a little bit more. Take a look."

Huge screens roll down from the stage, and the story of the attack, pulled together from news reports, plays in front of the audience. I start to feel like I'm floating in a dream. I sense a hand tugging my elbow, and a volunteer in one of my blue shirts whispers, "He wants you up there." I rise and slowly make my way past the front row of people whose gazes are glued to the video. In the dark, I am led to a big curtain onstage. I can't see the screen, but I hear someone's voice from an old interview.

"I honestly thought I was dying, and I feel very fortunate to be alive today." The voice is raw agony. It came from the burn unit. It belonged to me.

The video ends, the lights come up, and all eight thousand people jump to their feet. Their applause sounds like a rainstorm. Brandi and Kendra are in the front row bawling their eyes out, and so am I.

I walk straight over to Phil and whisper, "I can't believe you did this." That's all I can muster. Phil puts his arm around me, and then gently raises my left arm like we've won.

The main reason I don't want to bring attention to my situation is that I'm not interested in shocking anyone or making them feel uncomfortable. I am somewhat concerned about how people who don't know me react to my gnarled face and body. They're on the outside looking in, and it's a pretty miserable sight. But now I'm on the inside looking out, and the view from up on that stage is amazing. Thousands of people clapping, sounding aloud the compassion in their hearts, sending it out to me and my family.

God, I wish all my daughters were here to see this.

And just as that thought passes through my mind, here come Shelli, Steven, and Angela, holding Kennedy's hand, down the center aisle toward the stage with tears in their eyes. All I can do is hold my hands over my mouth, like a newly crowned Miss America. Everyone from Phil down to Kennedy has pulled off the biggest surprise of my life.

I start to whimper like I do when I can't cry any harder.

"Well," Phil explains to the crowd, "she's just seen the rest of her family."

Utter pride and happiness fill every bit of my being. My girls are all crying tears that say, *We're so happy for you, we're so proud of you, we love you, till death do us part, Mom.* They share a look of awe and excitement and adoration. All my life, until I die, I won't ever forget the look on their faces. The girls are clapping with the standing ovation wrapped around them, and little Kennedy jumps up and down in pure excitement. Then Shelli lifts her onto the stage.

"These are all of Cindi's daughters who came down here without her knowing. But I mean literally, this is her now fresh out of surgery. She faces a whole lot of surgery. But Cindi has volunteered for every event I've ever done. She says, 'Hey, let me come and do it. You're doing something for America, and I'm going to be there.' So somehow she's here."

I am still here, I think, looking out at all these people, who are still standing and applauding.

"This is why terrorism will not work in America! This is why you cannot break the spirit of America!"

I nod as Phil speaks and give Kennedy a reassuring pat on her shoulder. She seems perfectly at ease staring out at the audience of thousands, but I am shaking so much that I'm afraid my legs will buckle.

"You all saw Cindi's car. The acid came through the windshield, ate through her, ate through the seat, the floorboard, and ate through the frame of the car, and absolutely destroyed it. And as a woman who worked three jobs, when you're without a car, it's a little bit of a problem."

I don't want to believe where Phil is going with this story, but he keeps harping on about my needing a car. *He's not possibly talking about giving me a car.*

Phil faces me. "But some of your friends from Oklahoma and Texas decided you ought not to be walking. So they have a surprise for you right over here."

My stomach drops to the orchestra pit. *Oh my God.* As Phil gently turns me to look to the left, my shaking hands cover my face again, and I almost forget to breathe. The crowd pounds their palms together and hoots and hollers even louder.

"Well, sit down over there so she can see it," Phil asks a group of people standing closest to a brand-new white Chevy Trailblazer.

I almost fall to my knees. It is absolutely the most queen-for-a-day moment I've ever experienced.

"So lady, you got a new ride and we are humbled by your courage." Phil gives me a hug, I tell him I love him, and a volunteer ushers me over to my new car.

Some of the volunteers tell me how excited Phil had been backstage before the event, but it can't compare to how I feel. I'm shocked, honored, and shaken to my core by one of the nicest things that's ever happened to me. It makes me want to give back for all that has been given to me my whole life.

Standing next to my new car, a producer from *Entertainment Tonight* asks me for a comment.

I say what I determine I will tell everyone from now on. "One random act of violence changed my life forever, so on my behalf, commit a random act of kindness."

Ever since this "grassroots old girl" started that personal campaign, many people have told me, "I'm a better person because of you." Now that's humbling. Phil has preached all along that every life is significant, and each of us needs to get passionate about that. There is something to every life. Beyond my family, I am beginning to find my something.

Even without the car, even before the Dallas workshop, I'd always said I'd do anything for Phil. I think his work is that important. But if he'd asked me, in advance, to stand up in front of thousands of people, watch a movie of the attack on my life, and accept the gift of a car, I could not have done it. Phil baptized me, albeit by fire, into the world of sharing my story with thousands of people and witnessing the lights in their hearts glow for a stranger. He also helped show me that terrible events don't have to have terrible outcomes. I can't say I'm glad I was attacked with acid, but this life-changing experience is bringing out a person in me I didn't think I could ever be, a person I enjoy. Life-altering crimes don't have to mean life-altering for the worse. Up to this point in my life, I'm not sure I've ever felt better.

But not everyone is drawn to the spotlight. Jim sits this one out for two reasons. After our last visit to the burn unit, he came down with shingles and was bedridden for a few days. It was the one time he was willing to admit serious pain—he said shingles hurt worse than his burns. But besides not feeling well, Jim shies away from any public attention. He doesn't accompany me on any of these outings, where he undoubtedly would always be recognized as the hero who drove us to safety and saved as much of our skin as he possibly could. While I expand my horizons and find healing through it, Jim draws his horns in, as he likes to say. He prefers nobody associate the attack on our lives with him. People ask him all the time, "Hey, weren't you the guy with Dr. Phil's sister-in-law?" He answers, "Yeah, I was," but he never gives any more than that, unless I ask him to. Maybe it's a man versus woman thing. He thinks if he doesn't talk about it, it'll go away; I find sharing it with as many people as possible lifts some of the burden.

Once the event ends, the police officer at Phil's workshop makes his way over to me.

"I should be thanking you," he says.

Just as he cracks a smile, I see the opening of his heart.

⁓

NOT EVERY ASPECT OF publicity is a pleasant experience. Several tabloid papers had written about the acid attack, getting many of the facts wrong and fabricating nearly all of the quotes. No surprise, since even the ones that quoted me had never bothered to talk to me!

"Who the hell is Gloria Gaynor?" I ask, as I scan a tabloid article about the attack.

"She's the singer who did 'I Will Survive,'" Shelli explains.

"And what song is that?"

"Mom, you're hopeless!"

"Well, according to this, I'm supposed to know it by heart." Shelli looks over my shoulder as I read aloud. "It says here that Phil told me, 'Don't give in to the pain. Fight! Remember the words of your favorite song—"I Will Survive."'"

The only problem is, it's pure fiction. I was never interviewed by any of those people. I'd not only never heard of the song, but Phil never said anything of the sort to me. He had offered to fight *for* me, so I could heal. Plus, he hasn't become world famous by borrowing advice from pop singers.

" 'She collapsed when she saw her ravaged face in the mirror,' " Shelli reads on.

"That's just a flat-out lie!" I yell.

I couldn't even see the first day in the hospital, much less stand. And when I finally did see my face, I didn't fall down—I laughed.

" 'Broaddus was thinking of suicide when the phone rang.' "

I can overlook most of the lies, but this one really upsets me, because nothing could have been further from the truth. Suggesting that I wanted to take my life was beyond insulting. It never crossed my mind. I was thrilled to be alive!

Clearly these national tabloids are out to sell anything but the truth. I've been trying so hard to do the right thing, to handle the attack the right way, doing what I thought would help my daughters, family, friends, and anyone facing a terrible crisis. It disturbs me that the tabloids distorted the story and printed the wrong way to handle a crisis next to my picture.

Fortunately, when I was in the hospital, I didn't see a lot of that trash. People saved the articles for me, and I read them later, but they didn't truly hurt me. People who know me know I'm not like that. Fortunately, I think most people know that trash belongs in the trash. I just don't know how that kind of crap can be published. What a waste of space. If any of them had just called to interview me, I would have told them the true story, that there is hope no matter what happens.

By contrast, the local media were a delight to work with. Every single one of them, from Oklahoma City to Wichita Falls, Texas, seemed to want the best for me, to focus on finding the person who committed the crime, to tell the story of Phil's reward and one woman's struggle. Every few months, they'd try to update the story, checking to see if the OSBI was close to an arrest and whether or not I cared very much that they weren't.

After one interview, a young female reporter told me, "I'll pray for you."

"That's the best thing you can do," I told her. It seems sad sometimes how God is taken out of everything for fear of offending someone. It's nice to know someone is praying for you. But even good intentions can be a little overwhelming.

After so much media coverage, people stop me and my girls in public. Most times they have wonderful things to say to us that make us feel so proud. But one afternoon, while I was having lunch with my girls, a woman approaches our table.

"Are you Cindi Broaddus?"

"Yes, I am."

"Well, I've been praying for you," she says, and before I can thank her, she drops to her knees and asks if I'll join her in prayer. I barely nod before she begins praying out loud in a near trance that lasts no fewer than fifteen minutes. Everyone in the restaurant looks at us. The girls and I feel terribly uncomfortable, but she takes no notice. Our meals become cold; the other patrons' stares even colder. It begins to look like this woman is doing this not for our good, but for hers.

The woman is squeezing my sore arms, which are bandaged and ballooned. The skin on my chest and left arm is being loosened using tissue expanders—large balloons implanted under the skin with a port. Two times a week, Angela fills them with saline solution. My surgeon says it will take three months to fully stretch the growing skin, during which time the expanders will swell—agonizingly—to the size of cantaloupes under my skin. Once they are removed, surgeons will use the flabby skin left behind to replace scar tissue and to create mobility, especially around my neck and in the crook of my arm. I've already undergone several surgeries to implant the expanders. My girls tell me it pains them to see my body so deformed. I don't let them know that the balloons can be as painful as they are hideous. They only hurt for the first twenty-four hours after they're filled, when I bump them on something, or when someone grabs them, as is the case right now.

I finally decide to gently pry my arms away, which seems to break the woman's trance. She finishes, I thank her politely, and we return to our cold lunch plates.

Displays of affection are often better left to private times and places. But just as I can't control what the tabloids print, I can't stop a person from making a spectacle on my behalf. Allowing the spotlight to shine on me for a little while means I have to endure a less than pleasant side that inevitably comes with it.

⟫

A PRODUCER FROM the *Dr. Phil* show calls. I'm not surprised, because Phil had asked that my story be featured on one of the first episodes of his new TV show. I'd had requests from other national media, but I turned them down, wanting to save the story for Phil.

I immediately call the girls to let them know they have four days to get themselves ready. They are thrilled.

"We're going to be on the *Dr. Phil* show!" Brandi says, as though he's a movie star rather than her uncle. She can't contain her excitement when she talks to Kendra.

"That's great, good for you. But I'll tell you what, I wiped your mom's tush as much as any of you did over these past several months. Don't you think I ought to get an invitation, too?" Kendra laughs as she says it, but she's a little envious.

A producer and photographer from the show fly to Duncan the next day and turn my house into a makeshift television studio. One at a time, the producer interviews me and the girls, taking all day to do it. She collects photographs showing Jim and me before the attack, in the burn unit, and after a few surgeries, for a video version of the attack and aftermath. It is emotionally exhausting for all of us to re-live the morning I nearly died. I'm wiped out physically, too, recently having had surgery to remove a tissue expander that got infected.

Two weeks earlier, Angela had pointed out a black spot on my neck. She thought the skin cells were necrotizing, a word I'd never heard before. Once she explained it was similar to gangrene, I saw my doctor the next day.

Angela was right. We hadn't even filled the balloon yet, and the skin around the port was infected. My doctor immediately put me on a strong round of antibiotics and scheduled an emergency surgery to take it out.

Another producer calls me to let me know she has four airplane tickets; I ask if we can make it five. Jim comes to mind, because I'd like him to be there, but I know he won't want any part of it. I explain that Kendra is like a fourth daughter to me, and that she's been by my side every step of the way. The producer says an extra ticket is not a problem.

I call Kendra.

"Hey, Kendra, how about wiping my tush in L.A. for me?"

At first she giggles at the joke, but then, abruptly, her laughter turns to silence.

As the reality sets in, I hear Kendra start to cry. In a small way, it reminds me of the surprise Phil gave me down in Dallas. Kendra never expected anything and is honored to be by my side as a daughter. It feels wonderful to make her feel special.

A limousine shows up at my house at 6:00 A.M. on November 5, 2002, too early for us to get any attention from the town. Our driver takes us south to the Dallas–Fort Worth Airport. And so begins our royal journey to Hollywood, where we are all treated like queens. The show's hair and makeup artists fix us up, and we are downright giddy right up until the taping of the program.

"Disasters, moments of crisis, personal ground zeros"—Phil opens his show by reading those words on the teleprompter. "Today, incredible survivor stories. The tragedy and triumph of exceptional people who are shaking off their visited fate and want their lives back. I want to tell you, my family and I know firsthand the courage it takes to survive a moment of crisis because my first guest, Cindi, is my wife's sister. You may remember Cindi and the horrible moment that came when maniacs did the unthinkable in her life. Take a look."

At first, the show's producers wanted me to be sitting next to Phil when the video ends, but at the last minute, they ask me wait backstage.

"Phil wants to introduce you to the audience," a producer tells me.

While I wait for the introduction, I feel tears well up in my eyes. Even though I've lived the story and seen it redone many ways, I sense that I've come so far to face such a big moment.

"It's a terrible thing to happen to somebody. I was minding my own business—I wasn't doing anything." I watch myself say the words. At the time I said them, lying in the burn unit in Oklahoma City, my face was scorched from the fresh acid pour, not yet cut in more than a dozen surgeries to come.

Seeing myself after all that time makes me realize how much I've focused on getting better, rather than the attack itself, and I am proud of my progress. Phil jokes that by landing an interview on his new show I've made it to the top, and I believe he's right.

Phil's voice narrates the story on tape. "Just months after the attack, thousands gathered in Dallas to give Cindi a special gift." For the first time, I see the videotape of my reaction when Phil gave me the car. I start to shake a little again, remembering what a great moment it had been.

Phil continues, "Since then the road to recovery has been a rocky one. Today, over a year later, the memories of what happened are still fresh. But Cindi's remarkable spirit has soared."

Then the tape shows clips from the interviews the producer had done in Duncan.

"I haven't had a problem with being a down person. I give myself credit there. I can give up and be bitter or I can be happy I'm alive. I would never want my children or grandchildren to think I was a bitter old woman who gave up. Sometimes I wonder if there's something wrong with me that I have just chosen to be okay with it."

Phil's narration goes on. "Cindi readily admits she couldn't have gotten through this ordeal without the love of her three daughters and granddaughter. They've stood beside her in her pain and triumph."

Watching myself talk is nothing compared to watching my girls. The moment they appear on the tape, I start to lose it. When we were interviewed in my home, we had done it individually, behind closed doors, so I hadn't heard what the girls had said. We're such a happy family, full of laughter and focused on the positive, that we don't sit around and dwell on the emotional impact the attack has had on each of us. But here we are doing it in front of a nationwide audience.

Brandi's face comes up first in the video. Her voice shakes as she holds back tears describing our time together during the first hours in the emergency room. "I told her I loved her and that I was very glad she was alive. And I knew she was in a lot of pain, but we would take care of her, and I just kept telling her I loved her."

Shelli is next, and I can tell she'd been crying, because her nose is red, her eyes glassy, and she has the sweetest expression on her face. "I feel very blessed. Some of my friends wish they could have my mother. She makes me feel so proud."

Angela lips are trembling. Her words are measured very carefully, as though she wants to say something profound and real.

"When I look at my mother, I see past her scars. They are always a constant reminder of what happened, but I look further into her soul, and see she's a woman full of grace and dignity."

The girls have come such a long way from that first press conference in the hospital. During those first dreadful days, they talked about an act of violence they knew nothing about; now they talk about their mother, about whom they know everything. They've given voice to their hearts, and their words have never been so eloquent.

Then Kennedy, who has never shied away from the limelight, repeats what she understood from the moment she saw me in the burn unit. "She's still the same Grandma."

With that, the tape ends, and Phil goes ahead with the introduction.

"With me now is my extraordinary and beloved sister-in-law Cindi. Please welcome Cindi."

Just like the Dallas event, the people in the audience jump to their feet, stopping their applause only to wipe away tears. My girls are all in the front row again, proud as ever. It's a very touching moment for me, but for some reason I don't even understand, I start to blow kisses to the studio audience. *Who am I to blow kisses like I'm some Marilyn Monroe?* I hope the audience finds it more endearing than ridiculous.

"I think you've made some new friends. What do you think?" I hear Phil ask me over the applause. "Now don't hide your mouth."

"I'm trying to hide my jaw because it's shaking!" I answer back. "My latest surgery opened up my mouth, much to the chagrin of my coworkers," I quip and Phil smiles.

Talking to Phil feels very natural, even in such a staged environment. It isn't exactly like sitting in his kitchen with him, but it's close. The only part that's a little different is that he has to ask questions to which he already knows the answers.

"How do you get your mind around such a random act?"

"Well, I called on some lessons I learned from my brother-in-law very early in life . . ." Phil gives me a grateful pat, as if he'd planted my words and I'd dutifully spouted them back. But the reality is, our discussion is spontaneous. "You've always preached to me that we have choices. I made a choice and decided not to let this man hurt me a second time. He did this once, but he wasn't going to get by with it. I could wake up every day and be bitter and angry, or I could wake up and see these three beautiful people.

"I thought I was dying. But I had a calm peace. I had a message to get across. I begged Jim, something has happened, I'm dying, and you need to pull this car over because you need to tell my children how much I love them because I'm not ready to leave them."

My voice breaks as I repeat our conversation, and all over the studio I can see people dabbing their eyes with tissues. Phil turns to the camera, and he says something that I'll wear as a badge of honor the rest of my life.

"We're going to talk more about this with my sister-in-law, who is one of the biggest heroes in my life, when we come back."

While most people watching would think Phil just called for a commercial break, I am fixated on what he's just called me—his hero. Talk about a "changing day"—to use his terms—in my life. Here is a man I admire so much, a man who more often than not helps people discover their faults in order to lead them to better places, and he isn't trying to change me. He told the world that I am a hero to him, that I handled this crisis in my life the best way I could have. Ever since the attack I've tried to hold myself up as a positive role model for someone, but I did not dream it would be Phil. At this moment, I feel very special to him, very significant, in a way I have never felt before.

When the show continues, we talk about my fear of traveling under overpasses, to which Phil notes, "It's a phobia you are entitled to, by the way. I won't even dog on you about that one." He asks me to share the story about Deron's plan for a CableOne escort, and how I've managed to steer clear of the "victim attitude." And Phil does not miss an opportunity to ask anyone who knows anything about the crime to call the Oklahoma State Bureau of Investigation to help solve it.

"Cindi, we are hugely proud of you," Phil says, closing the interview. His audience is giving another ovation, and fortunately, I refrain from blowing any more kisses. All of it feels like a dream.

The national exposure from Phil's show only convinces me that this story might help other people see the positive side of a crisis. My story might teach someone that even acid can't scar your soul. Something terrible might actually lead to an extraordinary experience.

NOT EVERYONE HAS a brother-in-law who's as good at giving sound advice in tough situations. Being on Phil's show was certainly memorable, but what's helped me get through my ordeal is to draw on those life lessons he tries to teach. And I don't give him all the credit. I am a strong person of good character in my own right. Yet sometimes I look back and wonder, *How would I have done all this without Phil in my life?* While I'm proud of what I've done for myself, he's a major part of the reason I'm able to share my story with people outside my small community, people who might be inspired by it.

But even if I didn't have Phil in my life, I know I don't need access to millions to make a difference.

MONTH AFTER MONTH, I receive requests to speak to various groups—schools, hospitals, and civic organizations, for example. I don't prepare a speech, because I figure I'll sound a lot better speaking straight from the heart than from words written on paper. I never turn down a request, even though each time I feel like I have no idea what to say. I always trust that, with God's help, I'll share whatever the audience needs to hear, judging by their nodding heads and their furrowed brows, their smiles and their tears. I'll get a message across that someone in the audience needs to hear on that particular day, at that exact moment. Sometimes the faces in the crowd tell me I need to lighten the tone and recount a few more jokes or sweet stories about my grandchildren. Kennedy's red booboo story is guaranteed to bring down the house. Other times, the question-and-answer session after my talk lets me know they want to hear more about my struggle.

After a speech at a rehabilitative hospital in Plano, Texas, I take note of a woman walking toward me from the back of the room. Her face is sad and defeated, but she is definitely determined to make her way through the crowd. She grabs my hand.

"I finished my first round of chemotherapy today. It's been one of the hardest days of my life. What you said was exactly what I needed to hear."

I ask, "What kind of cancer?"

"Breast," the woman replies.

"I'll keep you in my prayers. You know that half the battle is believing you can do it." The woman holds on to my words and is not ready to let go of my hand. I can't help but think that my struggle pales in comparison to hers.

Ginny Robinson, the RN who arranged for me to speak, bounces over with great enthusiasm but waits patiently until everyone who wants a private word with me gets one.

"Oh, Cindi!" she beams. "Your talk was phenomenal! Everyone felt so privileged to be able to hear your amazing story! I've never had a speaker who inspired every single person in the room. Thank you!"

I'm not sure how to respond, except to thank *her* for letting me share a piece of myself. It's rewarding to know that something good can come from just talking about overcoming something terrible.

People often tell me that I'm an inspiration, and I'm always a little stunned when they do. I'm more inclined to accept that my *story* is inspiring, not me as an individual. Believing it's the story, rather than me, means anyone can come through a life-threatening and life-altering experience. I suppose it's easier to be angry than to forgive, to stay stuck than to move forward, to wither than to bloom. But here people stand in amazement that I choose to take a positive road. Sure it's harder, but who wants to live the rest of their life as if it's already gone? If something out of the ordinary happens, I don't think you have to be extraordinary to survive it. You just have to believe you will get over it and make choices that show you believe.

"You must think you can," I always say. "Don't say you can't."

After the Plano speech, I'm almost as exhausted as if I'd had another surgery. Reliving the details of the attack, and then listening to other peoples' struggles, is emotionally and physically draining. I'd made the five-hour drive to Plano myself, not thinking of the toll a long road trip takes on a woman in her fifties. I suppose travel is always tiring, no matter your age or condition, when the shower, the bed, and the night sounds are not your own. But talking to this group was a healing experience for me.

This is the end of the nightmare. This brings some meaning to my suffering. This is one place where I find my peace.

⌒

FEW PEOPLE ARE given the opportunity to know their true colors, to live to see how they responded to the face of death. I've been given that gift. I have a privileged view of myself facing a dire situation, and I continue to witness the depth of my strength, my character, and my connection to the people I love. Now that I've survived, now that I've made choices to heal, I have one crossroads left to travel: what to do with the rest of my life. I know in my heart I have to create meaning out of my suffering. As Phil has said many times: to whom much is given, much is expected. I feel it is my duty to share my story. What I've learned won't help anyone if I hide away, lock the front door, and throw away the key. I know that's what many victims of terrible crimes choose to do.

~

I RECEIVE AN invitation to a Christmas party to be given for a burn survivor group at Baptist Hospital. The dressing changes for my burns ended months ago. I don't really want to return to the hospital unless I absolutely have to, but the invitation says the group might disband if more people don't show an interest.

I call Jim. "You know, it's been awhile, why don't we go up to that Christmas gathering at the hospital?"

Jim agrees. "This is as good a time as any."

The two of us make the trip that we'd made a million times before. In a way, it feels good to be walking up here, relatively healthy, for positive reasons: to celebrate life, to give thanks this holiday, and to *not* need the bandages on our burns changed.

The party is set up in the family room across from the nurses' desk. I had walked into that room so many times; at first for exercise, since it was about fifty yards from my room, then to visit with loved ones when I needed a change of setting.

"Hi, everyone! Merry Christmas!" I chirp as I walk into the room. No one answers back. All sorts of goodies are piled onto a long table in the center of the room, and everyone is seated in chairs pushed up against the walls. It's more awkward than a middle-school dance. At least they played music back then.

Jim and I don't know anyone except for the social worker, Anita, who acknowledges my cheery welcome and tries to get the party going.

"Okay. I guess we'll get started with the meeting," she says. *Meeting? I thought this was a Christmas party. Oh boy.*

She continues. "Why don't we go around the room? Each of you can introduce yourself, and tell how you were burned."

From the moment the first person opens his mouth, it's a party all right. A pity party.

"My name is John. I got burned sixteen years ago. It ruined my life. I don't go out anymore. And this is a real rough time of year for me and my family."

"My name is Pete. I was burned ten years ago in a motorcycle accident. I rarely get out because I can't stand the stares."

"My name is Judy . . ." And on and on.

Now, I'm not making fun of these people or begrudging any of them their despair. Everyone handles difficult life changes in different ways. But I've been to funerals more uplifting, and as far as I can tell, none of us is dead yet. Ironically, most of us have escaped death. What a triumph! But I nearly cry listening to all the sad stories of ruined lives. Finally, it's my turn.

"I'm Cindi. I was burned a year and a half ago in a random act of violence. Three weeks after I left the burn unit, I went back to work, and made 95 percent of my sales goal. People stare at me, but most times I don't care. If they really fixate on me, I just don't make eye contact. That's the way I deal with it. And I feel fortunate to get to hold my grandchildren every day—just to be alive, especially during this time of year. I give thanks to God for sparing my life."

Anita smiles. No one says anything. A few people stare at the ground.

Listening to the experiences of other *victims* reaffirms what I've already decided—that I have made a choice to let life go on—to be a survivor. I will not shut down. Even though people will stare, and I'll be a little afraid of facing them all, I'm not going to lock myself away. To me, no situation is uncomfortable, rude, or downright insulting enough to make me retreat into a cave forever.

⸺∽⸺

BESIDES, if I hide myself away, I'll miss the birth of my next grandbaby.

After her miscarriage, Shelli struggled to get pregnant again. On the very day she was scheduled for an appointment with a fertility expert, she took a home pregnancy test. It was positive.

In March, I was supposed to have surgery again to implant another balloon in my left arm. But once I heard the baby was due in April, I decided to delay it. I just didn't want to go under anesthesia for five hours and risk not coming out of it.

On the last morning in April, Shelli is going into labor.

Her mother-in-law and I are relegated to the hallway of the birthing center. My hands are sweating, my feet pacing, and I just can't stand the wait. I peek into Shelli's room.

"Mom, I don't care if you come in."

That's all the permission I need. I pop into the room, peer over Chad's shoulder, and moments later it seems Emma Jade pops right out.

"She's here!" Dr. Cox yells out. Shelli's labor is the easiest I've ever seen.

We all cry wonderful, glorious, happy tears. Emma is the cutest little newborn, with tiny, perfect features. My face hurts from smiling. To see Shelli so happy wipes away all the sadness we've gone through. Finally, I feel completely in the present. My ordeal seems more than two years past. Just like the night of Kennedy's T-ball game, all is right with the world. Watching my daughter hold her own daughter, I can't stop thinking, *She's waited so long for this. Dreams do come true!* Dreams for her and dreams for me.

I am a grandma again, and nothing will ever take that away from me.

9

Lawmen and Lawmakers

DETECTIVE SPARKS'S NUMBER flashes on the caller ID. In the beginning, I used to hear from him every two weeks. But four months have passed without a word. I wonder if he's calling to report on a lead, a suspect, or maybe even an arrest.

Jim and I have grown increasingly frustrated that the investigation seems to be leading nowhere. The OSBI doesn't have much to work with. They've found no witnesses. There were no fingerprints. The only physical evidence was my damaged car, acid on the roadway, and what the acid did to us. Since the day in the hospital that Sparks told us about the original seventeen-year-old suspect, he had discovered quite a bit more about the kid. The morning of the attack, he had argued with his parents and stormed out of the house. He had worked for a carpet cleaning company, and so had access to caustic chemicals. The angry young man was given a lie detector test about the incident, and he failed miserably. But apparently none of it was enough to make an arrest. There was no solid evidence linking him to the crime.

With my girls, Shelli, Brandi, and Angela, the day the Cindi Broaddus Act is signed into law

Detective Sparks quickly comes to the reason for his call.

"Cindi, next week the Newcastle police and I are getting together to pass out reward flyers for the two-year anniversary. This is really the last year we have to find your attacker before the statute of limitations runs out. Would you like to come along?"

"Of course," I say.

On June 5, 2003, Shelli drives with me up to Newcastle. We pull into the police station, and waiting for us are the chief, the chaplain, another officer, a new OSBI detective assigned to my case, and several reporters.

The chief greets me with a stack of flyers. Under the headline "$10,000 REWARD" is a brief description of the crime, a photograph of my damaged car, and an OSBI phone number.

One by one, the chief makes quick introductions. I've seen some of the reporters before, but no one in law enforcement looks familiar to me. As I shake their hands, everyone is friendly, but I notice how one officer in particular holds my gaze for a minute as if he expects me to say something directly to him.

"Nice to meet you," I say to him, just as I had to everyone else.

Chaplain Chilles Hutchinson turns to Shelli and me. "Would you like to ride in my car?"

"Do you have air-conditioning?"

He laughs. "Yes, ma'am."

"We'll take it," I say.

The chaplain's car leads the motorcade to the nearest strip mall. As he drives, I wonder what his role was the morning of the attack.

"Were you there that morning?" I ask.

"No," he pauses for a moment. "I just wanted to be here for this. But Officer Norman, who you just met back there, was the first one on the scene."

Oh my gosh. That was him. I have to see him again. Officer Norman must've been looking for some kind of acknowledgment from me.

"I wished I'd known that. The morning of the attack I was blinded and couldn't see anyone. I've never seen his face before."

We pull into the mall parking lot, and everyone regroups. Officer Norman's car is the last to park, so I excuse myself and head toward him. I have tears in my eyes.

I reach out to give him a hug. "I didn't know I was looking at the face of my hero. It was you, wasn't it?"

"Yes," he answers.

As we let go of each other, I look straight into his eyes. "Thank you so much for what you did. I am just in awe."

"No, I'm the one in awe of you. It is such a relief to see you in person. I didn't think I'd get the chance again. I have never been so happy to see somebody alive and thriving."

"You're part of the reason I'm here," I tell him, smiling. And then, I have to ask: "So that morning, were you the one I asked—?"

Before I can say it, Officer Norman finishes my question. "—am I naked?" His voice softens slightly. "Yes, but you were covered up by then."

It is a very touching moment for both of us. I can tell he's trying hard to hold back his tears.

Shelli walks over to us and gently touches my shoulder. "Mom, I think they're ready to start."

As we join the group, Shelli turns to Officer Norman. "What are the chances these flyers will help catch the guy?"

"I'm surprised it's taken this long. As terrible as the attack was, I thought we'd have an arrest the very same day. We were sure someone would start talking, that someone was going to brag. Believe me, I'd like to be the one to put him behind bars for a long time."

"We wouldn't mind that one bit," Shelli says.

As we all go from store to store, most of the managers are receptive to the reward signs. I have such a big entourage that I feel like I have the Secret Service escorting me from business to business.

After more than two hours, I'm getting tired. But I figure I can visit a few more businesses before calling it a day.

I enter a convenience store and start my spiel.

"My name is Cindi Broaddus. Someone threw acid from an overpass two years ago onto my car and hasn't been caught yet. I'm wondering if we could put these flyers up."

"You said this happened two years ago?" the girl asks.

"Yes."

"Did you go to the travel plaza?"

"Not today."

"No, I mean when it happened."

"Yes, that's where my friend pulled over and called for help."

The girl pauses for a moment. "My mother was one of the two ladies working up front. She called 911." The girl tells me that when Jim ran into the convenience store yelling, "Call 911, call 911!" her mother immediately reached for the phone, reacting instinctively to Jim's urgent plea and the sight of his own burned body. She said her mother would never forget the startling sound of a car door slamming, and the sight of me staggering around the parking lot.

"Please tell her I say thank you."

What a small world. It makes me think that surely someone in this tiny town knows something. *Maybe today is the day we'll find him. These people were a part of the first day of my story. Maybe this is where it will end.*

It seems strange that the attack made an impression on the girl that day—she even had a personal connection to it—but that after all this time, she had forgotten about it. She and her mother and everyone else in town have gone on with their lives, just as they should. But I have been fighting nonstop, first for my life, and then for ways to prevent it from happening to anyone else.

As the day ends, I have another word with Officer Norman. "I want you to know, while you've been fighting for me, I've been fighting for you."

~⌇

THE DAY DAISY LAWLER walks into my office at CableOne, she introduces herself as a candidate for state senate.

"I'd like to talk with you about some political advertising. I'm running for office."

I recognize Daisy immediately. She and her husband had come into my office years ago asking about advertising for his political campaign. Now, it's her turn. She has recently retired from teaching to run for the very same office he had held.

She inquires about availability of time for her ads on various cable networks we have in our market. As we visit and I answer her questions, my mind wanders. I can't believe my good fortune. I have already begun formulating a plan of my own, and I know I will need the help of my state senator.

As Daisy gets up to leave, I tell her, "I don't know who will win, but whichever of you does, you're going to see a lot of me at the capitol. I plan on lobbying for lights and coverings on every overpass in Oklahoma."

She replies, "If I win, I will welcome the challenge."

A few months later, I get a call from Senator Lawler, who has just begun to settle into office. "When are we going to get to work?" she asks.

I'm shocked that she's called me so soon after the election. She had no way of knowing, but she did not even get my vote. I liked the incumbent senator better, thinking she was more experienced for the job. But Senator Lawler drafts a bill and goes for it. By the end of the legislative session, the Oklahoma senate is debating a bill that would make it a felony to throw objects that cause injury or property damage off an overpass or bridge. Instead of a slap on the wrist with a misdemeanor, the crime would be punishable by up to ten years in prison.

Before she heads out to the senate floor, Daisy gives me something of an apology. "Cindi, I know this isn't everything you want. This is just a starting point."

"At least it's a starting point," I tell her. She's right. It's not as far-reaching as I'd hoped, but at least the punishment will better fit the crime. At least *something*'s being done. I believe God put Daisy in my path for this purpose, and she'll have my vote for as long as she's running and as long as I'm alive.

In the gallery above the state senate, my granddaughter, my girls, and I take our seats. As I look down, I see Daisy along the back row of mahogany desks, where the junior senators are seated. She's surrounded by tall marble pillars, and I think what a grand, stately place this is. Daisy looks up to find where we are sitting. As our eyes make contact, I smile, and she nods as if to say, "This is your day." It seems unbelievable that we are here, on the verge of making a mark on the state law books. Another chance to bring meaning to suffering.

I turn to my granddaughter. "Do you know what laws are, Kennedy?"

She shakes her head, so I explain why we're here today. I don't tell her about all the meetings I'd already had with legislators from the senate and the house, how they banged out most of the details in committee. How one house member even asked, "What, are we going to punish somebody for throwing a gum wrapper on the road?" And how I thought that kind of opposition would mean my bill would never see this day. I will never forget them asking me to stand and tell my story, knowing I'd have to convince this subcommittee in order to get it to the full house. That was one of the most important speeches I would ever give. My heart pounded in my chest. My mouth was dry. But my deep sense of conviction took control. I looked them right in the eyes and said, "Don't let anybody else suffer what we've suffered. Help me make the roads a safer place for everyone to travel." That's an argument even Kennedy can understand.

Then the head of the senate, the president pro tem, calls everyone to order. Daisy had explained to my kids that he's called the pro tem because he presides over the senate in the absence of the lieutenant governor. After a few minutes of discussion and votes on other items, I hear the pro tem recognize Senator Lawler.

"This bill is for one of my constituents in Duncan. You've all heard her story. The bill that I'm introducing today will strengthen the penalty for throwing any object from a bridge or an overpass."

And with that, a debate ensues. I'm not sure what's happening. I try to follow the legislators, their voices clear, loud, and quick. They're discussing whether they can name the new piece of legislation after me. Finally, they decide on a title: The Cindi Broaddus Act. I can't believe how far I've come, from the third floor of the burn center to the hallowed floor of our state senate.

They come to a vote so fast, I'm a little confused. The pro tem sounds more like a cattle auctioneer, and I'm not sure if he'd bought, sold, or taken the whole thing off the block.

"Any debate? Seeing none . . ."

I notice the electronic board above his head. Names are lighting up, and votes are being tallied.

He continues, "All ayes . . . the ayes in . . . all nays . . . the ayes have it . . . NEXT!"

"Oh my God . . ." I whisper.

Thrilled and disbelieving at once, I glance at the girls, instinctively cover my mouth with my hands, and try to hold back tears.

The vote is unanimous: 41 to 0. We've won. One step closer to making sure no one else will scream the way I did that horrifying June morning.

I KNOW THAT a change in the law will not protect everyone. But I have to do whatever I can, because some things still trigger an overpowering fear that someone I love will be hurt. The first time it happens, I'm not driving on the road; I'm standing in my kitchen listening to Kendra talk about her upcoming vacation.

I know she's excited to see her mom and dad. As I'm cutting vegetables for a dinner salad, I ask her when she's flying out.

She hesitates for a second, which causes me to stop what I'm doing and look at her. "Tomorrow at six," she says reluctantly, her eyes averting my gaze.

"Please tell me it's at six P.M."

"Actually, Cindi, it's six A.M. But I'll be fine. Don't worry about me."

I suddenly feel a surge of terror seize my body. I can't believe Kendra's words. The answer I fear most is the one she gives. Through tears I ask her how she could have booked such an early departure, so close to the time Jim's and my flight was scheduled to leave.

Kendra tries to reassure me again. It takes me a minute to realize that my reaction is over the top, and I'm shocked at myself. Eventually, I calm down and apologize to Kendra, and the next morning she safely makes her way to the airport and beyond.

I'm not proud of the way I scared Kendra, but I hadn't realized how frightened I still am.

Fortunately, this kind of near panic attack happens only one other time.

~

IN SEPTEMBER, Jim, my brother-in-law Wayne, his son-in-law Bryce, and I are driving to Irving, on our way to Texas Stadium. I love football, and the only team I love more than the Oklahoma Sooners is the Dallas Cowboys. Thanks to the thrill of an impending game against the Buffalo Bills, I don't think much about the one-hundred-and-fifty-mile road trip. But my carefree spirit doesn't last long.

I'm in possession of a red parking pass, which I think means I get to park right up close. *A red VIP parking pass. We're hot stuff. This is good.*

I'm sorely mistaken. VIP obviously stands for "very inconvenient place." My family and I have to park all the way on the other side of the interstate, and then hoof it nearly a mile to the stadium. As we start on the long haul, I notice we have to do something I haven't done in years: walk across I-35 on an uncovered bridge.

"I'd rather walk across eight lanes of highway traffic than walk over that bridge. I'm serious. I don't know if I can do this." I think I say the words aloud, but no one seems to pay any attention to me, so they must be stuck in my throat. My family is probably nervous for me, too, but at this point we don't have any clear alternatives, so no one says a word as we walk.

We are surrounded by hundreds of fans all heading in the same direction. Suddenly I feel as though I am walking alone, and my pace slows. My family moves ahead, swept forward by the crowd. I don't like the feeling of being on an overpass with so much vulnerability passing under me: defenseless cars, innocent drivers, and sleepy passengers. Unintentionally, I drift over to the edge of the bridge and hold the railing with my hands. I am shaking. I want to look over, look down to see what my attacker saw. My knees turn to Jell-O as my gaze is drawn to the highway below. The view is frightening yet mesmerizing. My stomach starts to feel sick, and I start to feel dizzy.

I imagine how fast the glass jar of acid must have fallen, so quick it looked like a flutter to Jim. I expect to hear a scream.

"Cindi!" I know the voice. I recognize the urgency. It is Jim.

"Get away from the edge!" He grabs my arm and drags me away from the railing.

I still feel like I'm in a trance. "Oh my God. That's the closest I've ever come to knowing what he must've felt like."

"Come on. You're okay," Jim says as we walk to the stadium and find our seats.

⌐

MY COWBOYS WIN, putting all of us in high spirits. By this time, it's pitch-black out, and the crowd exiting the stadium is bigger than the one that ushered us in. My nephew-in-law, Bryce, is taller than the rest of us shrimps, and so we just follow his bobbing head. There are so many people that we are all shoulder to shoulder. Someone moos out loud as though we are cattle in a crowded pen.

At the foot of the overpass that we have to cross again, I turn to Jim. "I've become such a worrywart. I hope that thing was built to support this much weight."

The bridge is filled wall to wall with people, and all I can imagine is the overpass collapsing, spilling thousands of people onto the interstate.

Jim teases, "I do, too, 'cause we're fixin' to be right in the middle of it."

Suddenly, my panic intensifies. "Hey, this isn't the right way." We're still being pushed forward.

Now I'm shouting. "Guys, this isn't the right bridge!"

Jim looks at me like I'm insane. "What?"

"This is the wrong overpass."

"How do you know?"

"This has fencing shields on either side of it, and the one we crossed the first time was wide open."

Bryce trusts that I know what I'm talking about. "I'm sure you're right, Cindi, but there's no turning back."

We continue to walk over the partially covered bridge and march all over creation to find our car. This time, I don't mind the very inconvenient parking. The trip away from the stadium was not as stressful as the one going toward it. I don't have to think about it very long to know why—the bridge's fencing made all the difference. It takes one trip across an uncovered overpass to reinforce my belief that what Daisy and I are working for is very important.

During the car ride home, I find myself flashing my high beams at every overpass. I have to know if someone is up there. I know, realistically, that the chances of anything being dropped onto me and my car again are probably one trillion to one.

"Sorry about that," I say under my breath.

"Cindi, who are you talking to?" Jim asks sleepily from the passenger seat.

"All the people I'm blinding with my headlights. I know they can't hear me, but I think they'd understand why I have to see the top of every single bridge."

"I wish they would just find the guy, so you don't have to worry about him at all," Jim says.

It is frightening to think that he is free to hurt another person again. He dreamed up this devastating scheme. What else could he possibly come up with? I hope that day never comes, but it's possible he's already struck again.

~

TWO MONTHS AFTER Officer Gary Norman helped me pass out reward flyers, he calls me with some terrible news—someone targeted him and his family with an eerily similar weapon: an acid bomb in his mailbox.

He tells me that, just as it was for Jim and me, it was so early in the morning that the sky was still dark. Officer Norman walked out on his front lawn to pick up the paper. He wasn't fifty feet from his mailbox when it exploded.

The homemade bomb sprayed screws and acid from a glass bottle. Fortunately for Officer Norman, he was far enough away from it that he was untouched. Fortunate, too, that his wife and two children weren't yet waiting for the school bus at the end of the driveway next to the mailbox.

Immediately, Officer Norman retrieved a flashlight and his gun from inside his house.

"My first thought was, it's somebody I've arrested who just got out of jail."

He shined the flashlight on the broken pieces of glass and small screws scattered around his yard. Around the corner, he saw headlights of a car driving slowly down his street.

"Returning to the scene of a crime is one of the more idiotic temptations of a criminal. So I'm thinking he's coming to assess the damage."

As the driver passed in front of Officer Norman's house, he quickly pointed his flashlight toward the car. The driver jammed down on the accelerator, but not before Officer Norman got a tag number, and a description of the kids inside.

"Even with all that I saw, we couldn't arrest anybody. I don't know if the crimes are connected or not."

I hope, for Gary's sake, that the attack was random, not targeted. I hate the idea that someone who has worked tirelessly to help bring my case to a close would be threatened. And knowing that his family was at risk is almost too much to bear.

All the more reason to fight to protect the innocent.

※

"THEY HAVE A LEAD, Cindi!" Robin practically shouts into the phone out of pure excitement. "They think they may know who attacked you," she says, nearly out of breath.

There have been days over the past two and a half years when I wondered if detectives had forgotten about my case, but never Robin and Phil. Besides upping the reward money, they'd hired their own private investigators to track this guy down. Phil made a promise with those yellow roses, that he would "find the bastard" who did this. I have total faith in Phil, but it doesn't mean that capture is possible. Random crimes are nearly impossible to solve. I can tell by Robin's voice, however, that she isn't about to give up.

"Isn't this wonderful news?" Robin asks.

My heart skips a beat and then begins doing double time.

"What happened?" I ask reluctantly.

"Some lady called the 800 number of Phil's show. She says she knows someone who knows who dropped the acid on you. Apparently this lady is in trouble with the law herself and wants to help your investigation to reduce her charge. I can't believe you're not more excited!"

As much as I appreciate my sister's high hopes, I don't share them. I won't let myself. Why would a credible lead come to the *Dr. Phil* show instead of the Newcastle police? Is it possible that some sicko simply made up the tip just to get some publicity?

It may sound strange, but this is not wonderful news to me. Even if they do have a credible lead, I'm not sure I am ready to face that reality. I had given up hope that the person who threw acid on me would ever be found, much less charged. Not to be pessimistic, I had simply shelved the idea, and any emotional response to it. The thought of an arrest, a trial, possibly taking the stand and reliving it all—I can't even stomach it.

Turns out, I do not have to. Just as I suspected, it turns out that the lead is a dead end.

IT'S A COOL October morning, and the change in seasons reminds me of how much time has gone by since the crime, and how much I've tried to keep it from happening again.

"Are we going to see the judge?" Kennedy asks me as we drive north.

"No, sweetie, we're going to see the governor."

Kennedy, Angela, and Shelli are driving to the capitol with me for the official signing ceremony for the Cindi Broaddus Act.

I look up as we drive under a covered overpass, a rare sight on the highway. It's more like a chain-link cage wrapped around a bridge that connects a soccer field to a neighborhood. In the middle of it, a young couple is holding a toddler and pointing down at the traffic. They're smiling and waving, so I smile and wave back. That's the way it should be.

Kendra and Brandi are waiting for us as we pull into the parking lot of the capitol.

"Have you been here long?" I ask. They shake their heads to say no. "More important, should I wear this sweater around my neck?" I ask. To anyone else, the question might seem ridiculous considering my embattled body and face. The grafting around my mouth has not taken as well as my doctors and I had hoped. Instead of matching seamlessly, the skin grafted from my stomach to my chin turned dark yellow. Even a few laser treatments, which one surgeon thought would lighten my skin tone, didn't help my face look any less like that of a ventriloquist's dummy. But my daughters have always taken pride in helping me look my best—scars and all—when it might matter. I happen to think getting my picture taken with the governor warrants a little fashion consultation.

I feel better than ever. My cheeks are flushed with excitement, my eyes are wide open to see it all. After two years, three months, and twenty-five days, my smile is back. The most recent surgery, as painful as it was, opened my mouth fully. It's an awesome day to smile again.

Senator Lawler, Kendra, Kennedy, my girls, and I line up behind Governor Brad Henry. He takes a seat at a long desk in front of us and looks up to address the TV cameras shooting the event.

"This is an important piece of legislation. It'll make the roads of Oklahoma safer," the governor declares.

As he signs the Cindi Broaddus Act, I think of how *this* moment is clearly not a random act. It is deliberate, hard-fought, and necessary for the good of my home state. It's also a great act of kindness that could help protect millions of people.

Shelli looks down at Kennedy who is peering over the governor's left shoulder. She's straining to see what he's doing.

"Do you remember what a law is, Kennedy?" Shelli asks.

Kennedy shakes her head.

This time, Shelli explains. "You know how if you break the law, like if you speed, the police will pull you over and you'll get in trouble? Well, now there is a Cindi Broaddus law, and it means if you throw things at cars, you'll get in big trouble."

"Like when that man threw gasoline on Grandma?" Kennedy may not remember all the details, but she knows this isn't just a field trip for fun.

"Exactly like that," Shelli assures her.

The governor puts down his pen and faces me. "Cindi, thank you for your perseverance."

"You are welcome, Governor Henry."

I hope he knows he'll see me again, because I'm not finished. The trip to the Cowboys' stadium reinforced it. I want overpasses to be lighted or covered or changed in some way to protect the public below them. I want my nightmare replaced by a dream—of making bridges safe places on which families can walk, and under which drivers can drive, without fear. It's an expensive idea, but what's the price of saving someone from a scarred life? Or saving them from death? What if it had been Kennedy's face instead of mine?

As Kennedy crawls into Governor Henry's lap for a photograph, I realize that I'm not fighting for me. I'm fighting for her. For all the sweet faces and innocent lives that need to be protected.

"You know, that's a piece of history right there." Brandi almost sounds like she's joking, but as we walk out of the state capitol together, we all feel some pride in knowing it's true. Before we get in our own cars to go home, we ask a passerby to take our picture in front of the capitol.

By the time we head out of Oklahoma City on the drive south to Duncan, night is falling. Angela sits up front with me on the ride home. Not long into the drive, Kennedy falls asleep in the backseat.

"Mom, did you see that?"

"See what, Angela?"

"That car in front of us. It flashed its lights at that overpass."

"Are you sure?" I hadn't noticed. But Angela is certain.

I take a good look at the gray car ahead of us. The wheelchair lift attached to its roof looks familiar.

"Angela, do you think that's Randy and Sharon Burum's car?"

They are my neighbors across the street. We speed up to confirm our suspicion, and we're right. Angela and I wave to them as we pass by, but I have a lump in my throat. I know they flashed that overpass because of what happened to me. It makes me so sad that they feel fear as they drive the road that leads to home.

We all turn in to our street about the same time. Angela and I pull up to their driveway. I roll down my window.

"Hey, thanks for flashing your lights at those overpasses for us!"

"Cindi, we just have to," Sharon smiles a little, but also shakes her head. "We just have to."

That sums up how I feel about fighting for these new laws, for better protection, for good people like Sharon and Randy. *I just have to.*

My birthday—celebrating life and random acts of kindness

10

Random Acts of Kindness

A LOT OF PEOPLE still ask if I ever look back and wish I'd done something differently that might have changed my fate. But I don't. What purpose does it serve? What if I had skipped Kennedy's T-ball game and gone along with Jim's plan to stay the night in the city to be closer to the airport in the morning? What if Jim had driven more quickly, more slowly, or in another lane? What if I'd canceled the trip because of Brandi's nightmare?

I've always believed that there are no accidents—that nothing happens without a reason. I know why I have lived through a random act of violence—to respond in kind with a thousand acts of kindness.

The moment after I was attacked, an outpouring of kindnesses began. They came from every person along every crossroads: from Jim's care to Officer Norman's efforts, from the ladies who called 911 to the surgeons and nurses at the burn center, and from friends and neighbors who cheered for me to perfect strangers who prayed for me.

People willing to do acts of kindness seem to be everywhere; they just need to be inspired. The owner of Ted's Cafe Escondido added three thousand dollars to Phil's reward money, and he didn't want any recognition. I saw kindnesses in many other forms—letters, cards, flowers, plants, prayers, smiles, hugs, even donations to a medical account from doctors' offices, the telephone company, banks, classrooms of kids, department stores, grocery stores, restaurant owners, an entire town ("Everyone in Sherman, TX, wishes you well" the note read), people I'd known when I was a child, and people I'd never met. Every time I looked out my window that first year after the attack, someone was mowing my lawn. I never asked anyone to do it, they just did. When I joked with a neighbor that I wished they'd stop scalping my lawn and do something really great like paint my house, the next time I looked out the window, she had arranged just that. And kind words are just as important as kind acts. The other day, my friend Jana told me, "I hope the person who did this to you reads your book. He will see what I see—that God has worked in your life. God is a part of your healing. God could do the same for him."

My faith is stronger because of what has happened to me, not weaker. I don't believe that God does mean things, or purposely lets them happen, or abandons people. I just happened to be in the wrong place at the wrong time. I doubt that the person who did this to me meant to burn me like this. As he dropped the glass jar of acid off an overpass onto a small, speeding car with tinted windows, he had a one in ten trillion chance of actually injuring me the way he did. Yes, he could've missed. But the damage could've been worse. I could've easily lost my sight. I could've died. There is no doubt in my mind that God was watching out for me. And He still is. God always provides when I feel tested. He prepared me for this, pulled me out of the valley of death, and now leads me to sunnier pastures every day. The message I take away is this: I've fought the battle. I am getting my reward. And I am duty bound to share its wealth with everyone who hears my story.

Every so often, I send an e-mail to all the employees of my company throughout the country to let them know how I'm doing. So many of them are people I've never met, and yet they continue to send encouraging messages to help my healing.

One afternoon, Brandi told me that she sent a blanket e-mail to everyone at her office, asking for a favor.

"Mom, I hope you don't mind, but everyone keeps asking what they can do for you and for us. I went ahead and asked all of them to commit an act of kindness on your behalf."

"Of course I don't mind. It occurred to me that I should do the same. After a few words about the status of the investigation in my next companywide e-mail, I typed:

If I could ask one more thing from each of you, it is that you say a prayer that this person will be caught so they can no longer hurt anyone else OR that they may have turned their life around for the better. I can never begin to thank each of you for everything you have given me through prayers and good wishes. I am a different person today than I was before, but a much more enriched one. If you can, please commit a random act of kindness not only for me but for all the people who need one.

Cindi Broaddus

The next day, I was standing in line at the grocery store when I felt someone walk up behind me and gently tap me on the shoulder. I turned around expecting to see someone I knew. Instead it was the friendly face of an older woman whom I had never met before.

"Honey, you don't know me, but I've been praying for you," she said.

"Thank you. It's the prayers of people like you that have helped me through this."

I have said this to many people, many times, but it still amazes me—every time—that a total stranger can find the kindness in her heart to pray for another stranger. This lady recognized me from the TV and newspaper pictures. Her little tap on my shoulder reminds me of how much good there is in the world. And I've found goodness in the smallest places. Sometimes it comes in the light touch and gentle words of a stranger. It always comes cuddling on a rocking chair with my granddaughter. Since the attack I've been blessed to hear it and feel it so many times.

Most of us go through life without realizing how much we can do in the most effortless ways. Though I've always considered myself a kind person, I don't know if, in the past, I was ever unselfish enough to pray for someone I didn't know. Do I do it now? All the time. Hopefully I, too, can do something good for the people I pray for. To me, that goodness in the world is evidence of the way God works in our lives.

I SEEK CHANCES to spread my message, taking advantage of every opening. Quite often the opportunity comes up with strangers in public places such as airports, restaurants, and stores. I either see someone looking at my scars, or the person comes right out and asks what happened to my face. After sharing a short version of the crime, I try to persuade my listener that I don't want their outrage or sympathy—I want them to do a kindness in my name.

Sometimes opportunity knocks in surprising places, like Kennedy's elementary school. All along, I've made myself visible to Kennedy's friends and schoolmates so that she will never feel uncomfortable with me by her side. One afternoon in September, Kennedy brought me to class for Grandparents' Day.

All the children were sitting at their small wooden desks. The grandparents' chairs were lined up against the wall. The seat I chose happened to be right across from a little boy named Jared. He had his head down on his desk as if he were sleeping.

The class began their program. As Jared lifted his head to turn back and look at the door, I noticed he had tears streaming down his cheeks.

I leaned over and whispered, "What's the matter, Jared?"

He sniffled. "My grandpa was supposed to come and he's not here."

"Well, I'm sure Kennedy won't mind sharing her grandma."

Jared shook his head. "No, that's okay." And put his head back down in his folded arms.

After a few minutes, Jared's grandfather walked through the door. Jared's face lit up as he wiped away his tears. It was as if his grandpa were the only person in the room.

After some singing, cookies, and punch, the teacher invited us to read to the kids on future Friday afternoons.

A few weeks later, it was my turn to read, and I showed up at the school.

After I read a short story to Kennedy's class, the teacher asked aloud, "Do you mind telling another story? The one about what happened to you?"

"I don't mind at all," I told her.

In words I felt that six-year-olds could understand, I tried to explain what had happened. But I also wanted to reassure the children that good is far more powerful than evil.

"Sometimes people do mean things. You need to remember that it hurts when people are mean. So I try to do acts of kindness. It makes me feel better to do them. You should try to be kind all the time, too. It makes the world a happier place."

Jared piped right up. "You mean like the day I thought my grandpa couldn't come, and you said Kennedy can share? What you did for me when I was sad, that was an act of kindness, right?"

I was stunned by Jared's question. I couldn't help but think, *Oh my gosh, this is going to work*. I had no idea that he would remember my offer. Besides, at the time, my words didn't seem to console him at all. I'm amazed that he made the connection between the lesson I shared that day, and the way I treated him weeks ago. I showed this sweet, six-year-old boy something intangible, and he got it.

"Yes, that's exactly right," I told him. "Thank you for saying that."

My hope is that he never forgets it. My hope is that he pays it forward.

THE TRUTH IS, as much as I want to teach children about kindness, I do most of my learning from my own girls.

Three of the best examples I can share happened in the year and a half after the accident.

The first was a complete surprise to me.

We have a tradition in our family. On the day of your birthday, you don't have to do anything you don't want to. You don't have to make your bed. You don't even have to shower. The birthday girl gets to name what she wants to eat and where. Over the years, I have always taken full advantage of this. I don't do a thing on my birthday. Years ago, my girls started calling me the Birthday Queen. One year I wore Angela's high school homecoming crown on my head for the entire day.

I doubted that my first birthday after the accident was going to be festive. I imagined I'd spend the day with the girls and their families doing something quiet at home. Since I was still recovering from the latest surgery, there was no place I'd rather be.

I was vegging on the couch when Angela called. I almost didn't want to get up to pick up the phone.

"Yeah?" I answered in a deliberately lazy voice.

"What's the plan for dinner tonight, Birthday Queen?"

"I don't know. Since you all are busy painting your house, why don't we just order in pizza?"

"Okay. Sounds good."

A few minutes later, Angela came over to my house.

"Mom, do you care if we go to Eduardo's instead?" Eduardo's is my favorite Mexican restaurant in Duncan. "Kennedy wants to hear them sing their silly birthday song to you."

Angela knew she'd hit my sweet spot. "Oh sure. We'll go down there."

A few hours later, everyone came to pick me up.

Joe, the owner of Eduardo's, greeted us and started to walk us toward the left dining room, which is reserved for special occasions.

"Surprise!"

The room was packed with people, and balloons of every shape, size, and color. All my friends, my coworkers, my neighbors, my family, Jim—everybody was there. I immediately started crying. As everyone took a seat, I made my way around the room, hugging and thanking each person.

When I stopped to visit with my next-door neighbor Jane, she asked me, "Cindi, have you seen the invitation?"

She handed one to me that had been lying on the table. "It's one of the most touching things I've ever read."

I looked for myself. It read:

SURPRISE PARTY

We're lucky she's alive, so thankful she's still here
We're lucky to be celebrating her birthday this year

She is strong, she is tender, she makes us all complete
She inspires us to hope in the face of defeat

Throughout our mother's healing, she may not look the same
But we recognize the heart attached to her name

Come and celebrate and help us decorate the room
A message tied to a string, at the top of it a balloon

Whatever it is you feel about her, funny or very sweet
When she reads your message it will be her birthday treat

For all that she is you could never put a price
Come celebrate her birthday but most of all her life!!

Jane was right. It was beyond touching. And it explained why there were so many balloons.

I walked over to a huge cluster of balloons with messages attached. Some simply said, "Your life is our gift." A note from Angela read, "Mom, someday when my children ask me what hero means, I will tell them . . . Grandma! I love you."

My girls and I had talked about committing acts of kindness to counteract the violent act done to me, but I never imagined they and everyone I cared about would embrace the idea like that. The heartfelt messages were far better than anything my friends and family could've wrapped up in a box.

Once everyone left, the girls and I stared at the fifty or so balloons. "What do you want to do with them, Mom?" Shelli asked.

"Let's get them home first. I'd like to save these messages," I told her.

It took two SUVs to get all the balloons home. There seemed to be even more of them once we hauled them into my dining room. I couldn't see my own ceiling.

I enjoyed their company for a few days, and then decided it was time for them to go.

Brandi asked if we should let them fly away with new messages. She suggested we take a planned act—the birthday surprise—and turn it into many random acts of kindness.

We all agreed it was a great idea. So we tied a handwritten message to the end of each balloon that said: "If you find this, please commit a random act of kindness."

The next morning, Brandi, Angela, and I loaded up Angela's SUV, each of us holding the balloons. A few of them had popped, and the rest had lost some air, so they almost fit inside. Brandi and I held the ones that didn't out the windows. We all ducked around the balloons, laughing at the way they kept batting us in the head.

I yelled out, "Drive slowly, Angela!"

We drove to Shelli's house to pick her up. We had planned to drive all over town to release the balloons, one at a time.

"I've got an idea," I said as we greeted Shelli in her driveway. "Let's let the first one go right here."

Suddenly our mood became serious as we realized our home-grown idea was about to take flight for the first time. We piled out of the car, making sure that none of the balloons escaped. I gently pulled one of them out, careful not to tangle its string.

I read aloud the note we'd written, and then opened my fingers, letting the balloon drift above our heads. Together, we all looked up to watch it fly away. It went about four feet and got stuck in a tree.

"Well, I can see this is going to do the squirrels a world of good," Brandi quipped.

We burst into laughter.

We decided to find an open field on the edge of town.

"I know where we can go," I told them.

This time when we pulled up, everyone grabbed a handful of balloons. We figured the wind would carry them in all directions. This time, the release was more like we had imagined. The balloons lifted up into the sky and out of sight. We smiled again, knowing full well they could end up stuck in someone's tree, but they could also fall—randomly—into the hands of strangers. For a moment I thought about the person who threw acid on me and hoped that he would find one. Or even better, that someone would commit a random act of kindness to him.

I turned to the girls. "You know, I would go through all of it all over again just to be right here with you."

I know that sounds crazy, but I feel privileged to have seen what I've seen—the best of humanity. I'm thankful for a journey that's shown me how one simple act of kindness is far more powerful than a random act of violence. Being given a second chance at life, I feel born again. My story has a brand-new beginning. I have never been so thankful to be alive. I have never had such clear conviction about what I must do for the rest of my life.

IN APRIL, Brandi and Angela decided to take their efforts to another level and showed me a second way to give selflessly. They had been so touched by all the kindness shown to me that they wanted to keep giving back. Brandi searched the Internet and found a challenge: The Avon Walk for Breast Cancer to be held in Dallas, Texas. The girls would have to run or walk twenty miles a day for three days, sleep in tents, and endure the heat, but they decided to go for it. They would walk in honor of three people—Angela's mother-in-law, a friend who'd had a double mastectomy in her twenties, and Jim's late wife, Joyce. And they'd do it on my behalf.

The girls started training and raised almost four thousand dollars before race time came around. Brandi even wrote a letter to Just for Feet, a sporting goods store, explaining my story, and the store outfitted them head to toe: two pairs of running shoes and shorts, shirts, and headbands.

During the racing days, I couldn't stop thinking about the girls. I hoped they were okay and not in over their heads. On the last day of the race, I was eager to go with Jim and Kennedy to the finish line at Southern Methodist University in Dallas.

As Jim drove, my mind wandered back over the long road we'd

traveled together and how it had changed us. Having the bond of a shared experience, we are more sensitive to each other's needs. We'll do anything to help each other. But the attack took a toll on our relationship. It separated us physically. In the first year after the attack, Jim found himself depressed and withdrawn. It seemed every time I recovered from the discomfort of an old surgery, I'd have to face the anxiety of a new one. Gone are the carefree days when we could pick up and go. Now we're both afraid of being on the road, and we worry when our kids travel, too. I get a sick feeling in the pit of my stomach thinking that I could get the same phone call from police that they did.

Jim and I have also been distracted for a very good reason—we are obsessed with our grandkids. Jim made sure to be there the moment his first grandchild, Brannon Maxwell Wiseman, was born. He joked with me, "It's a terrible thing to have to wait sixty-four years before getting a grandchild." After Brannon's birth, I noticed Jim's depression lift; he called me that very weekend to say how extraordinary it was watching his daughter become a parent. "It's like seeing your own child reborn, and, at the same time, getting to observe *your* child respond to their *own* child." Now he visits Laura, her husband, Brent, and Brannon down in Frisco, Texas, about once a month. Life always has a way of handing something good to you right after something bad.

Once Jim, Kennedy, and I arrived on campus, we made our way through a crowd of thousands of people and found a spot near the finish line. Just like Jim, whose wife had died of breast cancer, everyone around us had been affected by the disease.

A woman turned to me and asked, "Do you have skin cancer?"

I explained how I came to look like a skin cancer victim, and how my girls were racing to commit an act of kindness on my behalf. At first, the woman was appalled by the nature of the attack, but soon she was moved to tears to hear about the girls. I couldn't help but think, *These people are dealing with a deadly disease, yet somehow they find my story inspiring. Here they are giving strength to me, while my kids give hope to them. What a way to give back.*

It was so hot under the Texas sun that I worried about how difficult the race was for Brandi and Angela. The runners neared the finish line, but they weren't crossing it. They were waiting for the walkers to catch up so everyone could win together. I craned my neck to look for the girls.

"Can you see my mom?" Kennedy asked.

"Not yet, Tootie," I said. I wondered if I'd miss them in such a huge crowd.

Suddenly, I heard their voices.

"Mom! Mom!"

There were hundreds of moms around me, but I knew whose kids were calling. I looked through the pack of runners and walkers and saw Angela and Brandi giving me a thumbs-up as they made their way across the crowd to hug me. The girls were exhausted. They were covered in sweat, their faces bright red, their ponytails frayed, and their feet so blistered they were limping. Physically, they looked absolutely miserable. But I'd never seen such big smiles on their faces. They were so proud of themselves.

"Mom, guess what! We almost didn't get to walk. We hadn't raised enough money," Brandi said.

"But then some other walkers who were registering at the same time heard and pitched in the difference. Another random act, Mom!" Angela said.

Knowing they—and the strangers who helped them—did this for me was too touching for words. I threw my arms around both of them.

Before we left the campus, the girls ran over to a sign and asked Jim to take their picture by it. It read, "You have within you, right now, everything you need to deal with whatever the world can throw at you." Whatever the world can *throw* at you. A gallon-sized glass jar of sulfuric acid is no match for the good things we plan to throw back at the world.

~

MY DAUGHTERS SHOWED ME a third example of an act of kindness. Shelli already had an idea of how to commemorate the first anniversary of the attack. "What if we buy a tree in your honor and plant it in a park in Duncan?" she suggested. "That way it'll benefit everyone who lives here."

I loved the idea, and I was so touched by Shelli's thoughtfulness and amazed that my girls continue to find ways to carry on this mission.

Family and friends gathered in the park for the planting. My coworkers shut down the office for the afternoon. Our local paper covered the event.

With my girls and Kennedy by my side, we watched a gardener dig into the earth and place the tree in the hole. As he packed the dirt around it, I thought, *What a peaceful way to remember what happened to Jim and me.* Brandi stepped forward to say a few words. She unfolded a small white piece of paper. It was clear to me that she'd put a lot of thought into this day. She looked into my eyes as she began to speak.

"We are here today, not to focus completely on the random act of

violence committed one year ago today, but to celebrate the courageous woman who has surfaced since. All who know you know you
have always been strong when faced with adversity; happy when you
should've been down; and optimistic when the odds were clearly
stacked against you. But you have shown another side that will forever be embedded in our hearts and minds. And that is your remarkable ability to forgive even the most heinous of crimes. You possess
qualities so inspiring, I wish all humankind could experience them.
God has made you an example, and I can see why. For 365 days you
have been a positive role model for so many people who know you
and even for some who don't. Let today be an example to you that
you have the strength, attitude, optimism, courage, and heart to deal
with anything life might throw your way. You are truly an inspiration.

"We are planting this tree in honor of you today. Let the trunk be
the courage you have shown, the branches be the lives that you have
touched, and the leaves be the random acts of kindness inspired by
you or for you. Today begins a new chapter of your life, a life we are
all so blessed to be a part of."

What a beautiful metaphor. I had thought of the tree as a gift to
others, not even considering that the girls would see it as a symbol of
what I'd done and hoped to do. With a little tender care from nature, the tree will grow and change with the seasons, just like us. It
will branch out, reach toward the sun, and root itself more firmly
into its home.

After the planting, the manager of the nursery where the girls
had bought the tree approached us. He handed the girls their check.

"I'm going to go ahead and commit a random act of kindness
right now and give this back to you."

We were so moved by his gesture, that a simple "thank you"
didn't seem sufficient. But it was. That's how it works.

NOW THAT I make a conscious effort to spread goodwill, I find that random acts of kindness happen all around me.

Ever since the attack I've tried not to travel when it's dark out, but sometimes Mother Nature—and the airlines—get in the way of that plan. I am coming home from my first trip to New York City, and am I on a high. In September 2003, I'd gone on a whirlwind tour of some publishing houses in the Big Apple, and the experience had been invigorating. For the first time, a book about the random act of violence that changed my life and the lives of many others seems like it might come to fruition. The adrenaline had kept me going at a frenetic tempo, but as I take a seat at the gate, I'm wiped out.

A few delayed flights and missed connections later, my emotional, mental, and physical fatigue catches up with me. I practically collapse from exhaustion in the St. Louis airport, having given up on the idea of getting home before dark, much less getting home at all. Hurricane Isabel is within striking distance of the East Coast, and the threat alone is enough to throw flight schedules across the United States into a frenzy.

"You're Dr. Phil's sister-in-law!" I hear a woman shout.

"Yes, I am," I answer with a very tired and, I'm afraid, uninterested voice. Most people who make the connection want to know all about Phil, from what he wears to breakfast to whether he barks advice over the dinner table.

But this lady just wants to talk. We are both on permanent standby, it seems, so time is ours to fill. She wants to know how I am getting along, what I had been doing in New York, and when she can read my book. After about a half hour, the flight attendant calls out her name.

"Congratulations, you're going home," the flight attendant tells her. She won the last seat on the standby flight.

"Would it be all right if I gave my seat to someone else? I'd like this lady to take my place." Incredibly, the woman is pointing in my direction.

"No, no, please, I'll take the next flight," I protest.

"You're tired, Cindi, and you need to get home. I was already resigned to staying in a hotel anyway. You go," she insists.

"Thank you for your random act of kindness," I tell her. If I weren't so spent, I might've fought her a little harder, but she truly seems happy to have me take her seat. *People like this make the world wonderful,* I think. *Just like the tiniest candle in the blackest corner of the earth. I can't wait to write about this day, this moment, this kindness done to me when I least expected it.*

MOST ACTS OF KINDNESS shown to me or offered on my behalf are local, meaning they affect a small community of people. Sometimes, as in the case of the airplane seat, I'm the only one on the receiving end. But Phil is helping me change that. One February afternoon in his kitchen, he presents me with an opportunity to make a more wide-reaching impact.

Robin is making sandwiches for both of us when Phil comes in.

"Did you know I'm starting a charitable foundation, Cindi?" he asks. "The Dr. Phil Foundation."

"That's great," I answer him.

"I will have a board of trustees, and they'll allocate the money and make sure the cause is legitimate. And of course I want the money to go toward something that is close to my heart."

"Of course." I'm listening closely, not sure where Phil is going with this. He loves giving surprises more than anyone I know.

"One of the things that matters most to me is helping children who are grossly overweight and who have diabetes. My question to you is, would you be a trustee?"

Robin drops the knife she's been using to cut the bread. We look at each other in amazement.

Just to make sure I've heard him correctly, I ask, "You want *me* to sit as a trustee on *your* board?"

"Yes I do. There will probably be four trustees, and you're the first I've asked. There will be some travel involved, and I'm going to do a show about it, but I want to know if you're interested."

"Phil, I don't know what to say. I'd be honored to do it."

I can't help but feel that Phil is deliberately honoring the woman who has emerged since the accident—the one Brandi talked about in her remarks during the tree planting. He's presenting me with another chance to leave a mark on the world, a way to help lots of people who need it desperately. Of all the people Phil could ask to be on the board, he's asking me, someone who has only a high school education. Getting the chance to serve like this is another way I see purpose in the course my life has taken.

~

NEARLY THREE YEARS have passed since that fateful morning of the attack, and I have been struggling with one remaining complication as a result of my injuries—until today. Today I decide that I will not go through another surgery. I'm healed on the inside, and I accept that I'll never completely heal on the outside. I will wear the scars of some sad stranger's hatred for the rest of my life. After fifteen surgeries to repair my burned skin, I'm finished.

I did not plan for this decision, but I have reached my limit. It started this cool April morning around 9:00 A.M., when I tried to rub the soreness out of my left arm. I kept looking down at it, thinking my arm looked swollen.

Shelli stops by as I'm on my way out the door to go to work.

"Do you see anything on my arm? It's so sore."

"It just looks like a bruise under the skin, Mom."

"I must've slept on it wrong."

But within the hour, I am so sick, it's obvious something else is desperately wrong. It feels like the worst case of flu I've ever had. My muscles ache, my joints hurt, I can't stop throwing up, and the pain in my arm keeps getting worse.

The most recent surgery implanted a new tissue expander in my arm. Once it is filled, it will be tight, bulky, and uncomfortable, stretching my skin beyond what you might imagine the human body is capable of. That stretched skin will later be grafted to places like my neck so that I will have more mobility.

But this is not mere discomfort. By 10:00 A.M., I am in the ER and on morphine again.

The hospital in Duncan is just a few blocks from my house. It's where Angela works. As familiar as they are with me there, they are not familiar with tissue expanders, because the hospital does not have a burn center or specialize in plastic surgery. All they can do is manage my pain, try to bring down my 103-degree fever, and prepare me to move to Dallas. My arm is red, hot, and swollen to three times its normal size. Clearly the infection is spreading faster than the antibiotic can take control.

I don't say it out loud, but all I can think about in the ER is Dana. She was a friend of Angela's and also a nurse. One day she went home from work not feeling well, came down with a strep infection that moved into her blood, and within forty-eight hours was dead. I can't get thoughts of Dana out of my mind. That fear I felt three years ago in the burn center surfaces again. *Am I going to die?*

But then my thoughts turn to Taylor, the bravest little girl I've ever known.

Taylor was Kennedy's cousin and best friend. Five days before the attack on Jim and me, the girls were playing together when Angela and I suddenly noticed something terribly wrong.

As Taylor dug the dirt with her left hand, her right hand was hanging limply by her side. When Taylor stood up to walk across the yard, she dragged her right foot behind her. The right side of Taylor's body, from her eye to her foot, appeared to be paralyzed. The next day doctors discovered that she had an inoperable malignant tumor in her brain stem. She had only a few months left to live.

Taylor's death brought us the greatest sorrow we'd ever known. But she was the mightiest fighter I've ever seen. I never saw her cry; I never saw her give in. She made such a huge impact in so little time on her community that neighbors created a park in her name. On Christmas Eve, dozens of children gather in Taylor's park to sing carols to the heavens.

Now, remembering this four-year-old's incredible bravery in the face of circumstances far more dire than mine, I summon my own courage.

By the time Angela drives me to Zale Lipshay Hospital in Dallas,

I know what I am going to do. This is too much, and it isn't worth it anymore.

Dr. Rohrich, the surgeon who's handled all of my care, is out of town. When the surgeon covering for him walks into the room, I don't hesitate to tell him what I want.

"Take it out," I tell him.

"You've been through so much, and the expander is already in. I'd like to try to save it," he argues.

"Nope. I want you to take it out today."

"What if we wait one more night?" he tries to negotiate. "And if your arm is one hundred percent better tomorrow morning, we'll keep the expander in."

"Deal." It is an easy one to make because I know my arm won't be better in the morning.

When it isn't, he fulfills his promise, and on Sunday morning, I go in for what I determine will be my last surgery.

There was a time when I hoped surgeons could remake my face and chest, erase any sign of the acid burns. It seemed like the doctors who did the extreme makeovers on television should've been able to do the same for me, but burned skin has its limitations. Reconstructing burned skin is not like having a face-lift in which plastic surgery seems to be capable of fixing even the most heinous abnormalities. No matter how many surgeries I undergo, I'll always look like a burn patient. Even where my skin is supposed to be repaired, I have keloid scarring, which sometimes looks worse than the acid burn. The keloid scars are big and puffy and look like they need to be surgically removed themselves. Some of the grafting on my face didn't take well either. The area around my mouth is a shade of yellow that doesn't match the rest of my pink skin. My arms are heavily scarred. But after so many operations, the keloid scars, and discolored skin grafts, I'm through. It's not that I've given up on looking better one day. It's that I choose to accept how far I've come. *I am healed.*

Immediately after the expander is taken out, the infection disappears. Now that the pain is gone, there's nothing I want to do more than cuddle up close with Kennedy.

Home again, my granddaughter climbs up onto my lap and snuggles in. At seven years old, she doesn't fit as neatly as she did when she was a toddler. But I never tire of having her near me, and at least we no longer need a barrier of pillows between us.

She looks up at my face and runs her fingers lightly across the scars around my mouth. As they've been healing after each surgery, she hasn't paid much attention to them. It's ironic that now that my skin has been repaired as much as it can, she's become curious again. It must be difficult for her to understand why a three-year-old injury hasn't gone away.

"Grandma," she asks, "are you ever going to be normal again?"

I know she means to ask whether my skin will ever look unscarred again, but I can't help but laugh out loud. It takes me a few seconds before I can contain myself to answer her.

"No, Tootie Girl, this is as normal as Grandma gets!"

She doesn't look disappointed by my answer, probably because she sees right past the scars to the smile she put on my face.

Every survivor finds a new version of normal because we can never return everything to the way it was. Important things, like love for family, stay the same and often strengthen. My outward appearance, however, has changed significantly, and it is the biggest adjustment everyone has had to make. While it's clear that my granddaughter is still getting used to the way I look, I'm over it.

I am free.

ONE MONTH LATER, on Mother's Day, my girls hand me a gift. This one actually comes in a box. It's a hand-painted sign that reads HOME IS WHERE YOUR STORY BEGINS.

On the back of the sign is a simple message from them. "Thank you for an awesome start to our story."

I've always said my best accomplishments are Angela, Shelli, and Brandi, and after all I've been through, nothing is truer for me. Without them, I wouldn't have had the will to survive the accident, the burn unit, and the surgeries; the motivation to change a law; the honor of serving on a board of trustees of a charitable foundation; and the inspiration to write a book.

Their gift makes me think about how many times my life has had a new beginning: when God and my parents gave me breath, when my girls were born, and when their children came into the world. It also happened when I realized I had survived a brutal attack on my life. I now know that we have a choice to regenerate our spirits, renew our souls, and revisit our reasons for living. As this story about one random act of violence ends, my life story continues to bear witness to the countless acts of kindness born every day.

TODAY I GIVE THANKS

Today, I give thanks just for being alive.
For being able to work and live in a great city.

For the many friends who have made my ongoing recovery so much easier to bear.

Today, I give thanks for each employee of CableOne who has shared my pain and my triumphs.

Today, I give thanks for my three beautiful, caring, compassionate daughters who have never left my side and have walked through the pain with me.

Today, I give thanks for their families who shared them with me.

Today, I give thanks to the thousands of strangers who tell me I am in their daily prayers.

Today, I give thanks to Jim who was brave enough to keep driving.

Today, I give thanks that after meeting the worst of humanity, I have been fortunate enough to meet the best of humanity.

Today and every day, I give thanks.

Love,
Cindi Broaddus

Printed in the *Duncan Banner* on Thanksgiving Day, 2001.

Epilogue

IN JULY 2004, three years and one month after the attack, I get a phone call from Special Agent Lydia Williams with the OSBI. The case had been reassigned to her one year earlier. I wonder why she's calling.

She wants to tell me that the OSBI has to close my case. The statute of limitations ran out in June. She's sorry she wasn't able to charge anyone. Normally they have good evidence—DNA, hair, fibers, blood, and, at the very least, witnesses. They didn't have any of that.

"I understand," I say. I know this already.

Agent Williams describes how hard they tried to solve this case. They had had several suspects. One kid bragged he could make an acid bomb and destroy somebody with it. He wore all black, dog collars, and chains—the whole bit. Another kid was heavily into LSD, even got kicked out of school. But they could never get a confession.

I'm stunned by this new information. It takes me a minute to digest all of it.

Looking toward the future—my grandchildren Emma and Carson

"I know you tried your best," I tell her.

"I did, Cindi. We tried so hard. And we wanted to solve it for you and Jim. Your ability to forgive is something I'm not sure I have in me."

I am touched—and a little embarrassed—by Agent Williams's words. I can't imagine I'm different from any other victim. I think of all the cases she's handled and all the people who have suffered more than I have. But she tells me that the work my family and I have done in the public eye has made an impact. After people saw us on the news, *Dr. Phil,* and in person delivering reward flyers, the calls poured in. In the end, detectives interviewed an astounding 150 people regarding my case.

Agent Williams continues. "You can't imagine how many kids thought it was funny to throw things like Coke cans and rocks off overpasses until they heard about you and Jim. One girl told me that no one deserves that."

I ask her about the OSBI's best suspect, the seventeen-year-old who failed the lie detector test.

Agent Williams says he's one of several troubled kids who've cleaned themselves up. One's in college; another has a good job. "It seems they're trying to take the right road."

This almost takes my breath away. Turning their lives around *is* better than turning themselves in.

Agent Williams can hear that I've started to cry.

"I'm so sorry, Cindi, I don't mean to upset you."

I'm not upset. This moment—alone, in my kitchen, on an otherwise very ordinary day—is a tremendous release. I did not know how much I needed to hear this. But after all this time, my suffering suddenly has meaning—some good has come from such evil.

"Cindi, you've made a difference in people's lives—the detectives at the bureau, the police officers in Newcastle, those kids. I promise you that."

Agent Williams doesn't have to make this call. As far as I'm concerned, she has shown me one mighty act of kindness. I thank her for calling, hang up the phone, and let the tears fall.

My case is closed.

Acknowledgments

THIS BOOK would never have been possible without the tireless work of many people. First and foremost I am not sure I would have survived to write it had it not been for my daughters and their families. They put their lives on hold for more than three years to give me everything I needed to make it to this day. They have taken each and every step right beside me and when I could not go one more, they picked me up and carried me. Angela, Shelli, Brandi, Steven, Kennedy, Carson, and Emma, you are my heroes!

To J Max: It is hard to find the words to thank someone for saving your life. I owe you so much. From the moment it happened to this very day you have been my knight in shining armor!

To Kendra Sides: You were there from the very first moment. In my heart, you are my fourth daughter!

To Kimberly Lohman Suiters: I will always be in your debt. Your positive attitude made this journey so easy. You never allowed me to get discouraged. We cried together, we laughed together, but most of all we wrote together and I am extremely grateful for that opportunity. Tyler and Sloane, thank you for sharing her with me.

To Jan Miller: You are without a doubt the best literary agent in the country. You have given me an amazing gift by believing so strongly in me. You have such a huge heart and it is so full of love, energy, and compassion. Every time I leave your office or hang up the phone after talking to you I feel uplifted. Thank you for giving me the chance to tell my story.

To Jennifer Holder: I thank you for hours of work. This was a difficult task that you were handed and you did an awesome job.

To Phil McGraw: You are a great teacher, a great human being, and a great brother-in-law. Thank you for giving me the tools I needed to survive. You came to my rescue and gave my family and me the help we needed to make it through this. I will always love you for that.

To my sister Robin McGraw: You are gracious and caring. You gave me a soft place to fall. You know all the right things to say and you gave me a shoulder to cry on. Thank you for opening your home to me after each surgery where I had a warm and loving environment in which to heal.

To my sisters Jamie and Karin: Thank you for loving me and standing beside me throughout this entire ordeal.

To Roger, Wayne, and Jackie: Thanks for being the great brothers you are.

To the staff of Baptist Burn Center, I was lucky to be in your care.

To Dr. Rod Rohrich, Cheryl Koch, and all of your staff: thank you for giving me back my life. My faith and trust in you continues to grow.

To Tom Might, Ron Pancratz, and each and every CableOne Employee in Duncan and all over the United States: You are the greatest! You gave me time to heal while encouraging me every day with your cards, notes, e-mails, and contributions to the medical fund. Thank you for your patience, I could not work for a better company.

To Darrel Massie, Deron Lindsay, and Brian Bertrand: You dropped everything to rush to my side. Thank you for caring so much. You have always gone the extra mile for me; I will forever be grateful.

To Jennifer Brehl: Thank you for believing in my story. Your insight made my book so much better. You have an eye for detail that clearly shows what an exceptional editor you are. Thanks for being on my side.

To the Newscastle Police Department and Officer Gary Norman: You have a special place in my heart. I hold you in the highest esteem.

To Sharon Smith, Jenny Villagrana, Jane Ball, Klarice Dant, Brent and Laura Wiseman, Tim and Carrie Walker, Linda Burton, JD King, Buzz and Lee Ann Carter, John Galloway: Your friendship and help were invaluable to me. You did all the little yet so very important things that nobody saw and always did them from your heart.

To my mom and dad: Even though you are not here physically you have been the guardian angels on my shoulders. I miss you always and love you dearly.

To God: Thank you for sending these people into my life when I needed them the most and for loving me enough to make sure that I had the best mankind has to offer. Thank you for answering all of the prayers on my behalf.

To the people who continue to answer this random act of violence with a random act of kindness: You have made this journey worthwhile. Please continue . . .